4th International Hadron Physics Conference

(TROIA'14)

Journal of Physics: Conference Series Volume 562

Canakkale, Turkey
1 – 5 July 2014

Editors:

H. Dag
G. Erkol
A. Kucukarslan
A. Ozpineci

ISBN: 978-1-63439-666-0
ISSN: 1742-6588

Printed from e-media with permission by:

Curran Associates, Inc.
57 Morehouse Lane
Red Hook, NY 12571

Some format issues inherent in the e-media version may also appear in this print version.

Printed by Curran Associates, Inc. (2015)

Institute of Physics
Dirac House, Temple Back
Bristol BS1 6BE UK

Phone: 44 1 17 929 7481
Fax: 44 1 17 920 0979

techtracking@iop.org

Additional copies of this publication are available from:

Curran Associates, Inc.
57 Morehouse Lane
Red Hook, NY 12571 USA
Phone: 845-758-0400
Fax: 845-758-2634
Email: curran@proceedings.com
Web: www.proceedings.com

TABLE OF CONTENTS

Journal of Physics: Conference Series **562** (2014) 011001 doi:10.1088/1742-6596/562/1/011001

The 4th International Conference on Hadron Physics, TROIA'14, was held at Canakkale, Turkey on 1-5 July 2014. Ozyegin University, Middle East Technical University, Canakkale Onsekiz Mart University, Turkish Atomic Energy Authority and HadronPhysics2 Consortium sponsored the conference. It aimed at bringing together the experts and the young scientists working on experimental and theoretical hadron physics. About 50 participants from 10 countries attended the conference.

The topics covered included:

- Chiral Perturbation Theory
- QCD Sum Rules
- Effective Field Theory
- Exotic Hadrons
- Hadron Properties from Lattice QCD
- Experimental Results and Future Perspectives
- Hadronic Distribution Amplitudes

The conference presentations were organized such that the morning sessions contained invited talks and afternoon sessions were devoted to contributed talks. The speakers of the invited talks were: C. Alexandrou, A. Gal, L. Tolos, J.R. Pelaez and M. Schindler. We had also guest speakers D. A. Demir and T. Senger.

The conference venue was a resort hotel around Canakkale. As a social program, a guided full-day excursion to the excavation site of the ancient Troia town and Assos was organized.

We believe that this conference provided a medium for young scientists and experts in the field to effectively communicate and share ideas. We would like to express our sincere thanks to supporting agencies and to all participants for their contributions and stimulating discussions. We are also grateful to the Scientific Secretary, Bora Işıldak, and all other members of the Organizing Committee for their patience and efforts.

30.10.2014
The Editors

Hüseyin Dağ
Güray Erkol
Ayşe Küçükarslan
Altuğ Özpineci

Journal of Physics: Conference Series **562** (2014) 011001 doi:10.1088/1742-6596/562/1/011001

Peer review statement

All papers published in this volume of *Journal of Physics: Conference Series* have been peer reviewed through processes administered by the proceedings Editors. Reviews were conducted by expert referees to the professional and scientific standards expected of a proceedings journal published by IOP Publishing.

4th International Hadron Physics Conference (TROIA'14) IOP Publishing
Journal of Physics: Conference Series **562** (2014) 012001 doi:10.1088/1742-6596/562/1/012001

$N\Delta$ and $\Delta\Delta$ dibaryons

Avraham Gal

Racah Institute of Physics, The Hebrew University, 91904 Jerusalem, Israel

E-mail: avragal@savion.huji.ac.il

Abstract. Experimental evidence for $I(J^P)$=0(3^+) $\Delta\Delta$ dibaryon $\mathcal{D}_{03}(2370)$ has been presented recently by the WASA-at-COSY Collaboration. Here I review new hadronic-basis calculations of $L = 0$ nonstrange $N\Delta$ and $\Delta\Delta$ dibaryon candidates. In particular, $\mathcal{D}_{03}(2370)$ is generated dynamically in terms of long-range physics dominated by pions, nucleons and Δ's. These calculations are so far the only ones to reproduce the relatively small $\mathcal{D}_{03}(2370)$ width of 70-80 MeV. Predictions are also given for the location and width of $\mathcal{D}_{30}$, the $I(J^P)$=3(0^+) exotic partner of $\mathcal{D}_{03}(2370)$.

1. Introduction

The WASA-at-COSY Collaboration has presented recently striking evidence for a $I(J^P) = 0(3^+)$ $\Delta\Delta$ dibaryon some 80-90 MeV below the $\Delta\Delta$ threshold, with a relatively small width of $\Gamma \approx 70 - 80$ MeV, by observing a distinct resonance in the energy spectrum of $pn \to d\pi\pi$ reactions [1, 2] as shown in Fig. 1–left. Isospin $I = 0$ is uniquely fixed in this particular $\pi^0\pi^0$ production reaction and the spin-parity 3^+ assignment follows from the measured deuteron and pions angular distributions, assuming s-wave decaying $\Delta\Delta$ pair. The peak of the $M^2_{d\pi}$ distribution on the right panel at $\sqrt{s}$ ≈2.13 GeV, almost at the $\mathcal{D}_{12}(2150)$ $N\Delta$ dibaryon location (see below), suggests that $\mathcal{D}_{12}$ plays a role in forming the $\Delta\Delta$ dibaryon $\mathcal{D}_{03}$.

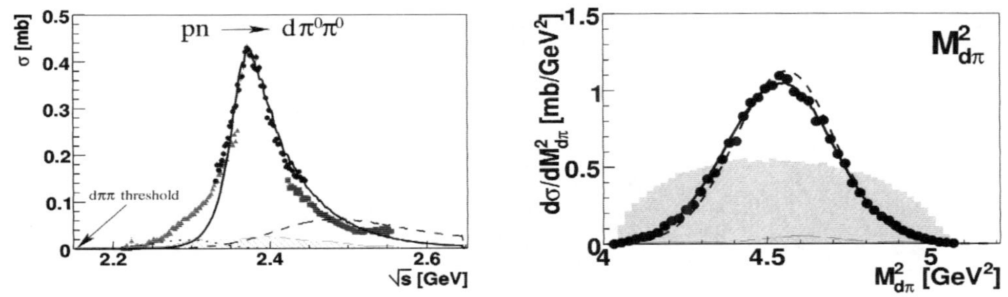

Figure 1. $\mathcal{D}_{03}(2370)$ $\Delta\Delta$ dibaryon resonance signal on the left panel, and its $M^2_{d\pi^0}$ Dalitz-plot projection on the right panel, from $pn \to d\pi^0\pi^0$ measurements by WASA-at-COSY [1]. This resonance was also observed consistently in $pn \to d\pi^+\pi^-$ measurements [2]. Figures courtesy of Heinz Clement.

Journal of Physics: Conference Series **562** (2014) 012001 doi:10.1088/1742-6596/562/1/012001

Further evidence supporting the $\mathcal{D}_{03}(2370)$ dibaryon assignment comes from very recent measurements of pn elastic scattering as a function of energy, taking sufficiently small steps around $\sqrt{s} = 2370$ MeV [3]. This is shown in Fig. 2–left for the Argand diagram of the $^3D_3$ partial wave, and in the right panel for the speed plot of the $^3D_3$ partial wave, within a new SAID partial wave analysis incorporating these measurements.

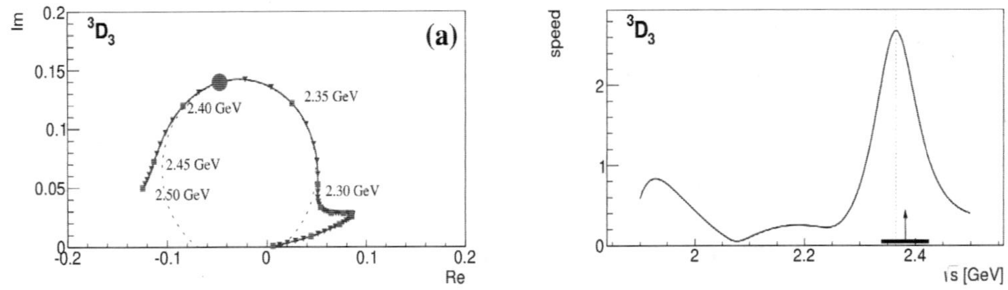

Figure 2. $\mathcal{D}_{03}(2370)$ $\Delta\Delta$ dibaryon resonance signals in the Argand diagram on the left panel, and in the speed plot on the right panel, both for the np $^3D_3$ partial wave, from recent np scattering measurements by WASA-at-COSY [3]. Figures courtesy of Heinz Clement.

$N\Delta$ and $\Delta\Delta$ s-wave dibaryon resonances $\mathcal{D}_{IS}$ with isospin I and spin S were proposed as early as 1964, when quarks were still perceived as merely mathematical entities, by Dyson and Xuong [4] who focused on the lowest-dimension SU(6) multiplet in the $\mathbf{56} \times \mathbf{56}$ product that contains the SU(3) $\overline{\mathbf{10}}$ and $\mathbf{27}$ multiplets in which the deuteron $\mathcal{D}_{01}$ and NN virtual state $\mathcal{D}_{10}$ are classified. This yields two dibaryon candidates, $\mathcal{D}_{12}$ ($N\Delta$) and $\mathcal{D}_{03}$ ($\Delta\Delta$) as listed in Table 1. Identifying the constant A in the resulting mass formula $M = A + B[I(I+1) + S(S+1) - 2]$ with the NN threshold mass 1878 MeV, a value $B \approx 47$ MeV was determined by assigning $\mathcal{D}_{12}$ to the $pp \leftrightarrow \pi^+ d$ resonance at $\sqrt{s} = 2160$ MeV (near the $N\Delta$ threshold) which was observed already during the 1950's. This led to the prediction $M(\mathcal{D}_{03})$=2350 MeV. The $\mathcal{D}_{03}$ dibaryon was the subject of many quark-based model calculations since 1980, see Refs. [5–13] for a representative although incomplete listing. Dibaryons were reviewed recently in Ref. [14].

Table 1. Nonstrange s-wave dibaryon SU(6) predictions [4].

dibaryon	I	S	SU(3)	legend	mass
$\mathcal{D}_{01}$	0	1	$\overline{\mathbf{10}}$	deuteron	A
$\mathcal{D}_{10}$	1	0	$\mathbf{27}$	nn	A
$\mathcal{D}_{12}$	1	2	$\mathbf{27}$	$N\Delta$	$A+6B$
$\mathcal{D}_{21}$	2	1	$\mathbf{35}$	$N\Delta$	$A+6B$
$\mathcal{D}_{03}$	0	3	$\overline{\mathbf{10}}$	$\Delta\Delta$	$A+10B$
$\mathcal{D}_{30}$	3	0	$\mathbf{28}$	$\Delta\Delta$	$A+10B$

It is shown below that the pion-assisted methodology applied recently by Gal and Garcilazo [15, 16] couples $\mathcal{D}_{12}$ and $\mathcal{D}_{03}$ dynamically in a perfectly natural way, the analogue of which has not emerged in quark-based models. Our hadronic-based calculations emphasize the long-range physics aspects of nonstrange dibaryons.

Journal of Physics: Conference Series **562** (2014) 012001 doi:10.1088/1742-6596/562/1/012001

2. Pion-assisted nonstrange dibaryons

The discussion in this section is divided into two subsections, the first one specializing to $N\Delta$ dibaryons and the second one highlighting the $\mathcal{D}_{03}$ $\Delta\Delta$ dibaryon.

2.1. $N\Delta$ dibaryons

The $\mathcal{D}_{12}$ dibaryon shows up experimentally as $NN(^1D_2) \leftrightarrow \pi d(^3P_2)$ coupled-channel resonance corresponding to a quasibound $N\Delta$ with mass $M \approx 2.15$ GeV, near the $N\Delta$ threshold, and width $\Gamma \approx 0.12$ GeV [17, 18] as shown in Fig. 3 for the Argand diagram of the $^1D_2$ partial wave in pp elastic scattering.

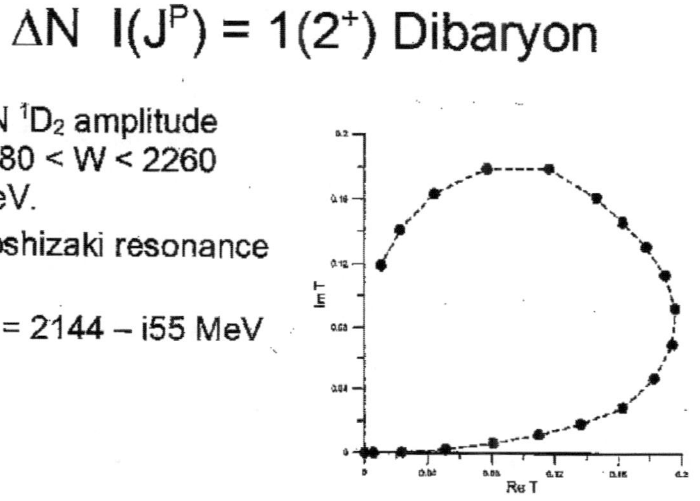

Figure 3. Argand diagram of the $^1D_2$ partial wave in pp elastic scattering from SAID, in agreement with past determinations of the $\mathcal{D}_{12}$ dibaryon resonance pole position, W=2148-i63 MeV [17] and W=2144-i55 MeV [18].

In our recent work [16] we have calculated this dibaryon and other $N\Delta$ dibaryon candidates such as $\mathcal{D}_{21}$ (see Table 1) by solving Faddeev equations with relativistic kinematics for the πNN three-body system, where the πN subsystem is dominated by the P_{33} $\Delta(1232)$ resonance channel and the NN subsystem is dominated by the $^3S_1$ and $^1S_0$ channels. The coupled Faddeev equations give rise then to an effective $N\Delta$ Lippmann-Schwinger (LS) equation for the three-body S-matrix pole, with energy-dependent kernels that incorporate spectator-hadron propagators, as shown diagrammatically in Fig. 4 where circles denote the $N\Delta$ T matrix.

Figure 4. $N\Delta$ dibaryon's Lippmann-Schwinger equation [16].

Of the four possible $L = 0$ $N\Delta$ dibaryon candidates $\mathcal{D}_{IS}$ with $IS = 12, 21, 11, 22$, the latter two do not provide resonant solutions. For $\mathcal{D}_{12}$, only ${}^3S_1$ contributes out of the two NN interactions, while for $\mathcal{D}_{21}$ only ${}^1S_0$ contributes. Since the ${}^3S_1$ interaction is the more attractive one, $\mathcal{D}_{12}$ lies below $\mathcal{D}_{21}$ as borne out by the calculated masses listed in Table 2 for two choices of the P_{33} interaction form factor corresponding to spatial sizes of 1.35 fm and 0.9 fm of the Δ isobar. The two dibaryons are found to be degenerate to within less than 20 MeV. The mass values calculated for $\mathcal{D}_{12}$ are reasonably close to the value $W = 2148 - \text{i}63$ MeV [17] and $W = 2144 - \text{i}55$ MeV [18] derived in coupled-channel phenomenological analyses.

Table 2. $N\Delta$ dibaryon S-matrix poles (in MeV) for $\mathcal{D}_{12}$ and $\mathcal{D}_{21}$, obtained by solving πNN Faddeev equations for two choices of the πN P_{33} form factor, with large (small) spatial size denoted $>$ ($<$).

$W^>(\mathcal{D}_{12})$	$W^>(\mathcal{D}_{21})$	$W^<(\mathcal{D}_{12})$	$W^<(\mathcal{D}_{21})$
2147−i60	2165−i64	2159−i70	2169−i69

2.2. $\Delta\Delta$ dibaryons

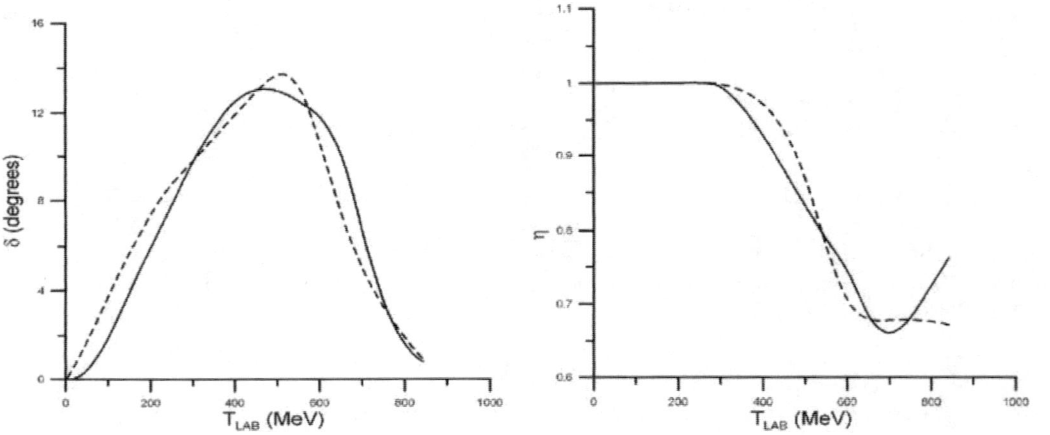

Figure 5. Coupled-channel fit (solid) to the SAID (dashed) NN ${}^1D_2$ phase shift δ (left panel) and inelasticity η (right panel), see text.

Four-body $\pi\pi NN$ calculations are required, strictly speaking, to discuss $\Delta\Delta$ dibaryons. In Ref. [15] we studied the $\mathcal{D}_{03}$ dibaryon by solving a $\pi N\Delta'$ three-body model, where Δ' is a stable $\Delta(1232)$ and the $N\Delta'$ interaction is dominated by the $\mathcal{D}_{12}$ dibaryon. The $I(J^P) = 1(2^+)$ $N\Delta'$ interaction was not assumed to resonate but, rather, it was fitted within a NN–πNN–$N\Delta'$ coupled-channel caricature model to the NN ${}^1D_2$ T-matrix, requiring that the resulting $N\Delta'$ separable-interaction form factor is representative of long-range physics, with momentum-space soft cutoff $\Lambda \lesssim 3$ fm^{-1}. A fit of this kind is shown in Fig. 5.

The Faddeev equations of the $\pi N\Delta'$ three-body model give rise, as before, to an effective LS equation for the $\Delta\Delta'$ S-matrix pole corresponding to $\mathcal{D}_{03}$. This LS equation is shown diagrammatically in Fig. 6, where D stands for the $\mathcal{D}_{12}$ dibaryon. The πN interaction was

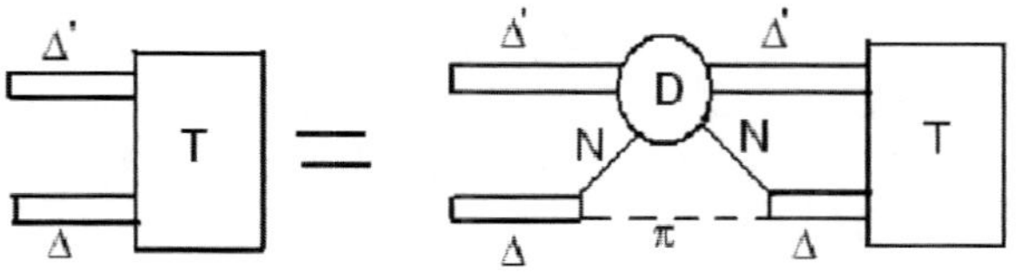

Figure 6. S-matrix pole equation for $\mathcal{D}_{03}(2370)$ $\Delta\Delta$ dibaryon [15].

assumed again to be dominated by the P_{33} Δ resonance, using two different parametrizations of its form factor that span a reasonable range of the Δ hadronic size. In Ref. [16] we have extended the calculation of $\mathcal{D}_{03}$ to other $\mathcal{D}_{IS}$ $\Delta\Delta$ dibaryon candidates, with D now standing for both $N\Delta$ dibaryons $\mathcal{D}_{12}$ and $\mathcal{D}_{21}$. Since $\mathcal{D}_{21}$ is almost degenerate with $\mathcal{D}_{12}$, and with no NN observables to constrain the input (I,S)=(2,1) $N\Delta'$ interaction, the latter was taken the same as for (I,S)=(1,2). The model dependence of this assumption is under study at present. The lowest and also narrowest $\Delta\Delta$ dibaryons found are $\mathcal{D}_{03}$ and $\mathcal{D}_{30}$.

Table 3. $\Delta\Delta$ dibaryon S-matrix poles (in MeV) obtained in Refs. [15,16] by using a spectator-Δ' complex mass $W(\Delta')$ (first column) in the propagator of the LS equation depicted in Fig. 6. The last two columns give calculated mass and width values averaged over those from the $>$ and $<$ columns, where $>$ and $<$ are defined in the caption of Table 2.

$W(\Delta')$	$W^>(\mathcal{D}_{03})$	$W^>(\mathcal{D}_{30})$	$W^<(\mathcal{D}_{03})$	$W^<(\mathcal{D}_{30})$	$W_{\rm av}(\mathcal{D}_{03})$	$W_{\rm av}(\mathcal{D}_{30})$
1211−i49.5	2383−i47	2412−i49	2342−i31	2370−i30	2363−i39	2391−i39
1211−i(2/3)49.5	2383−i41	2411−i41	2343−i24	2370−i22	2363−i33	2390−i32

Representative results for $\mathcal{D}_{03}$ and $\mathcal{D}_{30}$ are assembled in Table 3, where the calculated mass and width values listed in each row correspond to the value listed there of the spectator-Δ' complex mass $W(\Delta')$ used in the propagator of the LS equation shown in Fig. 6. The value of $W(\Delta')$ in the first row is that of the $\Delta(1232)$ S-matrix pole. It is implicitly assumed thereby that the decay $\Delta' \to N\pi$ proceeds independently of the $\Delta \to N\pi$ isobar decay. However, as pointed out in Ref. [15], care must be exercised to ensure that the decay nucleons and pions satisfy Fermi-Dirac and Bose-Einstein statistics requirements, respectively. Assuming $L = 0$ for the decay-nucleon pair, this leads to the suppression factor 2/3 depicted in the value of $W(\Delta')$ listed in the second row. It is seen that the widths obtained upon applying this width-suppression are only moderately smaller, by less than 15 MeV, than those calculated disregarding this quantum-statistics correlation.

The mass and width values calculated for $\mathcal{D}_{03}$ [15] agree very well with those determined by the WASA-at-COSY Collaboration [1–3], reproducing in particular the reported width value $\Gamma(\mathcal{D}_{03}) \approx 70$ MeV which is considerably below the phase-space estimate $\Gamma_\Delta \le \Gamma(\mathcal{D}_{03}) \le 2\Gamma_\Delta$, with $\Gamma_\Delta \approx 118$ MeV. No other calculation so far has succeeded to do that. Similarly small widths according to Table 3 hold for $\mathcal{D}_{30}$ which is located about 30 MeV above $\mathcal{D}_{03}$. This is about half of the spacing found very recently in the quark-based calculations of Ref. [13]. Note, however, that the widths calculated there are considerably larger than ours. A more complete discussion of these and of other $\mathcal{D}_{IS}$ $\Delta\Delta$ dibaryon candidates is found in Ref. [16].

3. Conclusion

It was shown how the 1964 Dyson-Xuong SU(6)-based classification and predictions of nonstrange dibaryons [4] are confirmed in our hadronic model of pion-assisted $N\Delta$ and $\Delta\Delta$ dibaryons [15, 16]. The input for dibaryon calculations in this model consists of nucleons, pions and Δ's, interacting via long-range pairwise interactions. These calculations reproduce the two nonstrange dibaryons established experimentally and phenomenologically so far, the $N\Delta$ dibaryon $\mathcal{D}_{12}$ [17, 18] and the $\Delta\Delta$ dibaryon $\mathcal{D}_{03}$ reported by WASA-at-COSY [1–3], predicting also an exotic $I = 2$ $N\Delta$ dibaryon $\mathcal{D}_{21}$ nearly degenerate with $\mathcal{D}_{12}$. We note that $\mathcal{D}_{12}$ provides in our $\pi N\Delta$ three-body model of $\mathcal{D}_{03}$ a two-body decay channel $\pi\mathcal{D}_{12}$ with threshold lower than $\Delta\Delta$. Our calculations are capable of dealing with other $\Delta\Delta$ dibaryon candidates [16], in particular the $I = 3$ exotic $\mathcal{D}_{30}$ highlighted recently by Bashkanov, Brodsky and Clement [19]. These authors emphasized the dominant role that six-quark hidden-color configurations might play in binding $\mathcal{D}_{03}$ and $\mathcal{D}_{30}$, but recent explicit quark-based calculations [13] find these configurations to play a marginal role, enhancing dibaryon binding by merely 15±5 MeV and reducing the dibaryon width from 175 to 150 MeV for $\mathcal{D}_{03}$, still twice as big as the reported width, and from 216 to 200 MeV for $\mathcal{D}_{30}$. Hidden-color considerations are naturally outside the scope of hadronic models and it is gratifying that the results presented here in the hadronic basis are independent of such poorly understood configurations.

Acknowledgments

Fruitful collaboration with Humberto Garcilazo and stimulating discussions with Mikhail Bashkanov and Heinz Clement are gratefully acknowledged. Special thanks are due to Güray Erkol and the other Organizers of TROIA14 for their kind hospitality.

References

[1] P. Adlarson et al. (WASA-at-COSY Collaboration), Phys. Rev. Lett. **106** (2011) 242302. See also the preceding reports: H. Clement et al. (CELSIUS-WASA Collaboration), Prog. Part. Nucl. Phys. **61** (2008) 276; M. Bashkanov et al. (CELSIUS/WASA Collaboration), Phys. Rev. Lett. **102** (2009) 052301.
[2] P. Adlarson et al. (WASA-at-COSY Collaboration), Phys. Lett. **721** (2013) 229.
[3] P. Adlarson et al. (WASA-at-COSY Collaboration, SAID Data Analysis Center), Phys. Rev. C **90** (2014) 035204. See also P. Adlarson et al. (WASA-at-COSY Collaboration, SAID Data Analysis Center), Phys. Rev. Lett. **112** (2014) 202301.
[4] F.J. Dyson, N.-H. Xuong, Phys. Rev. Lett. **13** (1964) 815.
[5] P.J. Mulders, A.T. Aerts, J.J. de Swart, Phys. Rev. D **21** (1980) 2653.
[6] M. Oka, K. Yazaki, Phys. Lett. B **90** (1980) 41.
[7] P.J. Mulders, A.W. Thomas, J. Phys. G **9** (1983) 1159.
[8] K. Maltman, Nucl. Phys. A **438** (1985) 669.
[9] T. Goldman, K. Maltman, G.J. Stephenson, K.E. Schmidt, F. Wang, Phys. Rev. C **39** (1989) 1889.
[10] X.Q. Yuan, Z.Y. Zhang, Y.W. Yu, P.N. Shen, Phys. Rev. C **60** (1999) 045203.
[11] R.D. Mota, A. Valcarce, F. Fernández, D.R. Entem, H. Garcilazo, Phys. Rev. C **65** (2002) 034006.
[12] J.L. Ping, H.X. Huang, H.R. Pang, F. Wang, C.W. Wong, Phys. Rev. C **79** (2009) 024001.
[13] H. Huang, J. Ping, F. Wang, Phys. Rev. C **89** (2014) 034001.
[14] A. Gal, in *From Nuclei to Stars, Festschrift in Honor of Gerald E Browm*, Ed. Sabine Lee (WS, 2011) pp. 157-170 (arXiv:1011.6322). See also M. Oka, Phys. Rev. D **38** (1988) 298.
[15] A. Gal, H. Garcilazo, Phys. Rev. Lett. **111** (2013) 172301.
[16] A. Gal, H. Garcilazo, Nucl. Phys. A **928** (2014) 73.
[17] R.A. Arndt, J.S. Hyslop III, L.D. Roper, Phys. Rev. D **35** (1987) 128.
[18] N. Hoshizaki, Phys. Rev. C **45** (1992) R1424, Prog. Theor. Phys. **89** (1993) 563.
[19] M. Bashkanov, S.J. Brodsky, H. Clement, Phys. Lett. B **727** (2013) 438. See also F. Huang, Z.Y. Zhang, P.N. Shen, W.L. Wang, arXiv:1408.0458 [nucl-th].

4th International Hadron Physics Conference (TROIA'14) IOP Publishing
Journal of Physics: Conference Series **562** (2014) 012002 doi:10.1088/1742-6596/562/1/012002

Renormalization and mixing in lattice QCD: The case of the chromomagnetic operator

M Constantinou[1], M Costa[1], R Frezzotti[2], V Lubicz[3], G Martinelli[4], D Meloni[3], H Panagopoulos[1,*] and S Simula[3]

[1] Department of Physics, University of Cyprus, CY-1678 Nicosia, Cyprus
[2] Dipartimento di Fisica, Università di Roma "Tor Vergata" and INFN, I-00133 Rome, Italy
[3] Dipartimento di Fisica, Università Roma Tre and INFN, I-00146 Rome, Italy
[4] SISSA, I-34136 Trieste, Italy
[*] Presented the talk. E-mail: haris@ucy.ac.cy

Abstract. We study matrix elements of the "chromomagnetic" operator on the lattice. This operator is contained in the strangeness-changing effective Hamiltonian which describes electroweak effects in the Standard Model and beyond. Having dimension 5, the chromomagnetic operator is characterized by a rich pattern of mixing with other operators of equal and lower dimensionality, including also non gauge invariant quantities; it is thus quite a challenge to extract from lattice simulations a clear signal for the hadronic matrix elements of this operator. We compute all relevant mixing coefficients to one loop in lattice perturbation theory; this necessitates calculating both 2-point (quark-antiquark) and 3-point (gluon-quark-antiquark) Green's functions at nonzero quark masses. We use the twisted mass lattice formulation, with Symanzik improved gluon action. We also provide a nonperturbative method to compute mixing with lower dimensional operators; we use this method in numerical simulations, and extract the mixing with the 3-dimensional scalar density, finding good agreement with one-loop results.

1. Introduction

The electroweak effective Hamiltonian describing strangeness changing ($\Delta S = 1$) processes, in the Standard Model (SM) and beyond, contains four "magnetic" operators of dimension 5:

$$H_{\rm eff}^{\Delta S=1,\ d=5} = \sum_{i=\pm}(C^i_\gamma Q^i_\gamma + C^i_g Q^i_g) + {\rm h.c.} \qquad (1)$$

$$Q^\pm_\gamma = \frac{Q_d\, e}{16\pi^2}\,(\bar s_L\, \sigma^{\mu\nu}\, F_{\mu\nu}\, d_R \pm \bar s_R\, \sigma^{\mu\nu}\, F_{\mu\nu}\, d_L)\,, \quad Q^\pm_g = \frac{g}{16\pi^2}\,(\bar s_L\, \sigma^{\mu\nu}\, G_{\mu\nu}\, d_R \pm \bar s_R\, \sigma^{\mu\nu}\, G_{\mu\nu}\, d_L)$$

The coefficients C^i_γ (C^i_g), multiplying the electromagnetic–EMO (chromomagnetic–CMO) operators, may be calculated perturbatively via the OPE; they are suppressed within the SM, but become more pronounced beyond the SM, e.g. through penguin diagrams in SUSY.

The matrix elements of the CMO are parameterized as (see [1] for a list of references):

$$\langle \pi^0 | Q^+_g | K^0\rangle = \frac{-11}{32\sqrt{2}\pi^2}\,\frac{M_K^2 (p_\pi\cdot p_K)}{m_s + m_d}\,B_{g1} \qquad (2)$$

$$\langle \pi^+\pi^- | Q^-_g | K^0\rangle = \frac{11\,{\rm i}}{32\pi^2}\,\frac{M_K^2\, M_\pi^2}{f_\pi\,(m_s+m_d)}\,B_{g2}\,, \quad \langle \pi^+\pi^+\pi^- | Q^+_g | K^+\rangle = \frac{-11}{16\pi^2}\,\frac{M_K^2\, M_\pi^2}{f_\pi^2\,(m_s+m_d)}\,B_{g3} \qquad (3)$$

These matrix elements are relevant for the study of $K^0 - \bar K^0$ mixing, ϵ'/ϵ, the $\Delta I = 1/2$ rule, and $K \to 3\pi$ decays. To leading order in χPT, the B-parameters are all related; thus, a lattice study of, say, Eq. (2), provides information for Eqs. (3) as well.

The EMO has been studied in simulations with $N_f = 0$, 2 dynamical flavors, focusing on $\langle \pi^0 | Q_\gamma^+ | K^0 \rangle$ which is relevant, e.g., for the branching ratio of $K_L \to \pi^0 e^+ e^-$ in SUSY models.

2. Operator Mixing – Lattice Action – Symmetries

A formidable issue in the study of the CMO is the fact that it mixes with a large number of other operators under renormalization. Even in dimensional regularization (DR), which has the simplest mixing pattern, the CMO ($\mathcal{O}_{CM} \equiv \mathcal{O}_1$) mixes with a total of 9 other operators ($\mathcal{O}_2 - \mathcal{O}_{10}$), forming a basis of dimension-5, Lorentz scalar operators with the same flavor content as the CMO. Among them, there are also gauge noninvariant operators ($\mathcal{O}_9, \mathcal{O}_{10}$); these are BRST invariant and vanish by the equations of motion, as required by renormalization theory.

$$\begin{aligned}
&\mathcal{O}_1 = g\,\overline{\psi}_s \sigma_{\mu\nu} G_{\mu\nu} \psi_d && \mathcal{O}_6 = \overline{\psi}_s(\overrightarrow{\not{D}} + m_d)^2 \psi_d + \overline{\psi}_s(-\overleftarrow{\not{D}} + m_s)^2 \psi_d \\
&\mathcal{O}_2 = (m_d^2 + m_s^2)\overline{\psi}_s \psi_d && \mathcal{O}_7 = m_s \overline{\psi}_s(\overrightarrow{\not{D}} + m_d)\psi_d + m_d \overline{\psi}_s(-\overleftarrow{\not{D}} + m_s)\psi_d \\
&\mathcal{O}_3 = m_d m_s \overline{\psi}_s \psi_d && \mathcal{O}_8 = m_d \overline{\psi}_s(\overrightarrow{\not{D}} + m_d)\psi_d + m_s \overline{\psi}_s(-\overleftarrow{\not{D}} + m_s)\psi_d \\
&\mathcal{O}_4 = \overline{\psi}_s \overleftarrow{D}_\mu \overrightarrow{D}_\mu \psi_d && \mathcal{O}_9 = \overline{\psi}_s \overleftarrow{\not{\partial}}(\overrightarrow{\not{D}} + m_d)\psi_d - \overline{\psi}_s(-\overleftarrow{\not{D}} + m_s)\overrightarrow{\not{\partial}}\psi_d \\
&\mathcal{O}_5 = \overline{\psi}_s(-\overleftarrow{\not{D}} + m_s)(\overrightarrow{\not{D}} + m_d)\psi_d && \mathcal{O}_{10} = \overline{\psi}_s \overrightarrow{\not{\partial}}(\overrightarrow{\not{D}} + m_d)\psi_d - \overline{\psi}_s(-\overleftarrow{\not{D}} + m_s)\overleftarrow{\not{\partial}}\psi_d
\end{aligned} \qquad (4)$$

On the lattice, where certain symmetries are violated, mixing can become considerably more complicated; there can be mixing with additional operators of dimension ≤ 5. To exclude as many as possible of these candidates, a judicious choice of lattice action is imperative.

We have adopted the twisted mass action for sea quarks and the Osterwalder - Seiler action for valence quarks. For gluons we have used the Symanzik improved action with several standard choices of values for the corresponding Symanzik coefficients. For our choice of action there appear only 3 additional operators, compared to DR, and they all have dimension < 5:

$$\mathcal{O}_{11} = i\overline{\psi}_s[r_d\gamma_5(\overrightarrow{\not{D}}+m_d)+r_s(-\overleftarrow{\not{D}}+m_s)\gamma_5]\psi_d\,,\ \mathcal{O}_{12} = i\,(r_d\,m_d+r_s\,m_s)\overline{\psi}_s\gamma_5\psi_d\,,\ \mathcal{O}_{13} = \overline{\psi}_s\,\psi_d \qquad (5)$$

3. Renormalization Matrix

Renormalized operators $\mathcal{O}_i^R$ are related to bare ones $\mathcal{O}_i$ via a 13×13 renormalization matrix Z:

$$\mathcal{O}_i = \sum_{j=1}^{13} Z_{ij}\mathcal{O}_j^R \qquad (\mathcal{O} = Z\mathcal{O}^R,\ \ \mathcal{O}^R = Z^{-1}\mathcal{O}) \qquad (6)$$

where $Z_{ij} = Z_{ij}^{X,Y}$ depend both on the regularization X ($X = L$ (lattice), DR (dimensional), etc.) and on the renormalization scheme Y ($Y = \overline{\text{MS}}$, RI', etc.). At tree level: $Z = \mathbb{1}$. For $\mathcal{O}_1^R$, we only need the first row of Z^{-1} (and thus, to one loop, only the first row or Z: Z_{1i}).

To obtain Z_{1i}, we have calculated, to one loop and in an arbitrary covariant gauge, the 2-point (quark-antiquark) and 3-point (quark-antiquark-gluon) bare amputated Green's functions of $\mathcal{O}_1$; these are related to the corresponding renormalized Green's functions through:

$$\begin{aligned}
&\langle \psi^R\,\mathcal{O}_1^R\,\overline{\psi}^R\rangle_{\text{amp}} = Z_\psi \textstyle\sum_{i=1}^{13}(Z^{-1})_{1i}\langle\psi\,\mathcal{O}_i\,\overline{\psi}\rangle_{\text{amp}}\,, && \psi = Z_\psi^{1/2}\,\psi^R \\
&\langle \psi^R\,\mathcal{O}_1^R\,\overline{\psi}^R A_\nu^R\rangle_{\text{amp}} = Z_\psi\,Z_A^{1/2} \textstyle\sum_{i=1}^{13}(Z^{-1})_{1i}\langle\psi\,\mathcal{O}_i\,\overline{\psi}\,A_\nu\rangle_{\text{amp}}\,, && A_\nu = Z_A^{1/2}\,A_\nu^R
\end{aligned} \qquad (7)$$

4. Results and Future Work

We calculated the bare Green's functions of Eq. (7), first in DR and then in the far more complicated case of the lattice. The purpose of the calculation in DR is twofold: First, it provides the coefficients $Z_{1i}^{DR,\overline{\text{MS}}}$; second, and most important, it leads to the $\overline{\text{MS}}$-renormalized Green's functions, which are then necessary for extracting the real quantities of interest: $Z_{1i}^{L,\overline{\text{MS}}}$.

An immediate, well-known by-product of $Z_{1,1}^{DR,\overline{\text{MS}}}$ is the anomalous dimension $\tilde{\gamma}_{CM}$ for the operator $\tilde{\mathcal{O}}_{CM} \equiv m\,\mathcal{O}_{CM}$: $\tilde{\gamma}_{CM} = g^2/(16\,\pi^2)\cdot(4\,N_c - 8/N_c)$. We find that the mixing coefficients for the gauge noninvariant operators $\mathcal{O}_9$, $\mathcal{O}_{10}$ do not vanish.

The lattice relations are formally again Eqs. (7); however all Z's now stand for $Z^{L,\overline{\text{MS}}}$, and the bare Green's functions must be calculated using the lattice regularization.

It is a highly nontrivial task to show that the lattice bare Green's functions, once renormalized, will coincide with those coming from DR: The $\overline{\rm MS}$-renormalized 3-point function has already an extremely complicated dependence on momenta and masses (involving Spence functions even for $m=0$), while the lattice bare 3-point function contains $\sim 10^5$ loop integrals depending on masses and external momenta. We have shown this equality in the most general case.

Our results, in the case of the Iwasaki gluon action, read ($\tilde{g}^2 \equiv g^2/(16\pi^2)$):

$$
\begin{aligned}
Z_{1,1}^{L,\overline{\rm MS}} &= 1+\tilde{g}^2\left(N_c(-7.9438+(1/2)\log(a^2\bar{\mu}^2))+N_c^{-1}(4.4851-(5/2)\log(a^2\bar{\mu}^2))\right), \qquad (8)\\
Z_{1,2}^{L,\overline{\rm MS}} &= \tilde{g}^2 C_F\,(4.5370+6\log(a^2\bar{\mu}^2)),\quad Z_{1,3}^{L,\overline{\rm MS}}=\mathcal{O}(\tilde{g}^4),\quad Z_{1,4}^{L,\overline{\rm MS}}=\mathcal{O}(\tilde{g}^4),\\
Z_{1,5}^{L,\overline{\rm MS}} &= \tilde{g}^2\left(N_c(4.2758-(3/2)\log(a^2\bar{\mu}^2))+N_c^{-1}(-3.7777+3\log(a^2\bar{\mu}^2))\right),\quad Z_{1,6}^{L,\overline{\rm MS}}=\mathcal{O}(\tilde{g}^4),\\
Z_{1,7}^{L,\overline{\rm MS}} &= -Z_{1,9}^{L,\overline{\rm MS}}=-Z_{5}^{L,\overline{\rm MS}}/2,\quad Z_{1,8}^{L,\overline{\rm MS}}=\tilde{g}^2C_F\,(-3.7760),\quad Z_{1,10}^{L,\overline{\rm MS}}=\tilde{g}^2C_F\,(3.7777-3\log(a^2\bar{\mu}^2)),\\
Z_{1,11}^{L,\overline{\rm MS}} &= a^{-1}\,\tilde{g}^2C_F\,(-3.2020)=-Z_{1,12}^{L,\overline{\rm MS}},\quad Z_{1,13}^{L,\overline{\rm MS}}=a^{-2}\tilde{g}^2C_F\,(36.0613)
\end{aligned}
$$

Preliminary non-perturbative results: In the calculation of on-shell matrix elements, by virtue of the equations of motion, some of the operators $\mathcal{O}_1-\mathcal{O}_{13}$ will not appear. The remaining ones: $\mathcal{O}_1$, $\mathcal{O}_2$, $\mathcal{O}_3$, $\mathcal{O}_4$, $\mathcal{O}_{12}$, $\mathcal{O}_{13}$ will be present, and it is crucial to have a stringent estimate of their mixing coefficients. For operators of the same dimensionality as the CMO, i.e. $\mathcal{O}_1-\mathcal{O}_4$, our one-loop results are expected to provide satisfactory accuracy; however, for operators of lower dimensionality ($\mathcal{O}_{12}$, $\mathcal{O}_{13}$), given that their coefficients are power divergent, perturbation theory is expected to provide only a ballpark estimate at best. Fortunately, it is precisely for the coefficients of these latter operators that we can have best access to non-perturbative estimates.

Imposing conditions such as:

$$\lim_{m_s,\,m_d\to 0}\langle\pi(0)|\mathcal{O}_1^{\rm sub}|K(0)\rangle=\lim_{m_s,\,m_d\to 0}\langle\pi(0)|\mathcal{O}_1+a^{-2}c_{13}\mathcal{O}_{13}|K(0)\rangle=0 \qquad (9)$$

$$\langle 0|\mathcal{O}_1^{\rm sub}|K(0)\rangle_{m_s,\,m_d}=\langle 0|\mathcal{O}_1+a^{-2}c_{13}\mathcal{O}_{13}+a^{-1}c_{12}\mathcal{O}_{12}|K(0)\rangle_{m_s,\,m_d}=0 \qquad (10)$$

we can fit the values of $c_{13}(g_0)$, $c_{12}(g_0)$ to data from simulations with varying quark masses.

In a preliminary series of simulations, we have extracted c_{13} at different values of the coupling ($\beta\equiv 6/g_0^2=1.90,\ 1.95,\ 2.10$).

The results for c_{13} exhibit a linear plus quadratic dependence on $\tilde{g}^2$, corresponding to a dominant one-loop effect (very close to the perturbative result) plus a subleading two-loop contribution, namely:

$$Z_{1,13}^{\rm non-pert}\sim a^{-2}\,\tilde{g}^2C_F\,35.8\,(1.0-3.0\,\tilde{g}^2) \qquad\qquad Z_{1,13}^{\rm pert}=a^{-2}\,\tilde{g}^2C_F\,(36.0613) \qquad (11)$$

Besides a series of controls which we have applied to our results, some further ones may be applied: (i) A calculation of 4-point Green's functions will provide important consistency checks, but no new information, on Z_{1i}. (On the other hand, n-point functions ($n>4$) are irrelevant: Being superficially convergent, they have a straightforward continuum limit.) (ii) Non-perturbative estimates of all mixing coefficients would be very important cross checks.

A further extension of the present work would be to apply methods of improved perturbation theory ("boosted" coupling, "cactus" diagrams, etc.) to our results.

Acknowledgments: The work of M. Constantinou and M. Costa was supported by RPF (Cyprus) under Contract No. TECHNOLOGY/ΘΕΠΙΣ/0311(BE)/16. D. Meloni acknowledges MIUR (Italy) for financial support under the program Futuro in Ricerca 2010 (RBFR10O36O).

[1] Constantinou M, Costa M, Frezzotti R, Lubicz V, Martinelli G, Meloni D, Panagopoulos H and Simula S 2014 *PoS* **(LATTICE 2013)** 316 (*Preprint* arXiv/1311.5057)

Journal of Physics: Conference Series **562** (2014) 012003 doi:10.1088/1742-6596/562/1/012003

Hadronic parity violation in few-nucleon systems

M R Schindler

Department of Physics and Astronomy, University of South Carolina, Columbia, SC, 29208, USA

E-mail: `mschindl@mailbox.sc.edu`

Abstract. The interactions between nucleons contain a parity-violating component, which originates in the weak interaction between quarks and which is suppressed by a factor of approximately 10^{-7} compared to the dominant parity-conserving component. A theoretical framework based on effective field theory methods to analyze and interpret parity-violating interactions between nucleons is described and a number of applications are discussed.

1. Introduction

Interactions between nucleons are the manifestations of the underlying interactions between the constituents of the nucleons. In addition to the parity-conserving (PC) strong and electromagnetic interactions, quarks also interact weakly. The interplay between weak quark-quark interactions and the strong interactions that confine the quarks into nucleons results in a parity-violating (PV) component in the nucleon-nucleon (NN) force. Because of the short range of the weak interaction of about 0.002 fm compared to the typical size of the nucleon of about 1 fm, the PV component of the NN interactions is sensitive to short-range correlations between quarks and can be considered a probe of nonperturbative QCD. However, the PV component of the NN interactions is typically suppressed by a factor of the order of 10^{-7}, making a detection of these effects challenging. It can be isolated by considering pseudo-scalar observables that would vanish if parity was conserved, such as longitudinal or angular asymmetries. For reviews, see, e.g., Refs. [1, 2, 3, 4].

Parity-violating effects can be enhanced by several orders of magnitude in heavier nuclei (see, e.g., Ref. [5]), but the analysis and interpretation of these results is complicated by nuclear structure effects. The ongoing developments in high-intensity neutron and photon sources and improved experimental control of systematics has made measurements of hadronic parity violation in few-nucleon systems possible. These systems can be analyzed in terms of two- and three-nucleon interactions without the complexities of nuclear structure. The following describes a program to analyze hadronic parity violation in few-nucleon systems that is based on effective field theory (EFT), and selected results are presented.

2. Hadronic parity violation in pionless effective field theory

Experiments aimed at PV observables typically involve very low energies much below the pion mass. At these energies, pion exchange between nucleons cannot be resolved. It is therefore possible to "integrate out" not just heavy hadrons, but also pions, and to formulate an EFT for nucleon interactions in which the only dynamical degrees of freedom are nucleons. The resulting "pionless EFT" or EFT($\not\pi$), consisting of contact terms with an increasing number of derivatives, has been very successful in describing low-energy observables, including electromagnetic effects. See, e.g., Refs. [6, 7, 8] for reviews.

In the PV sector, the leading-order (LO) Lagrangian contains five independent terms that describe the transition from two nucleons in an S-wave state to a P-wave state and vice versa [9, 10]. In a formulation that introduces an auxiliary dibaryon field for S-wave NN states, it is given by [11]

$$\begin{aligned}\mathcal{L}_{PV} = -\Bigg[& g^{(^3S_1-^1P_1)} d_t^{i\dagger}\left(N^T\sigma_2\tau_2\, i\overset{\leftrightarrow}{D}_i N\right) + g^{(^1S_0-^3P_0)}_{(\Delta I=0)} d_s^{A\dagger}\left(N^T\sigma_2\,\sigma_i\tau_2\tau_A\, i\overset{\leftrightarrow}{D}_i N\right)\\ &+ g^{(^1S_0-^3P_0)}_{(\Delta I=1)}\,\epsilon^{3AB}\, d_s^{A\dagger}\left(N^T\sigma_2\,\sigma_i\tau_2\tau^B\overset{\leftrightarrow}{D}_i N\right) + g^{(^1S_0-^3P_0)}_{(\Delta I=2)}\,\mathcal{I}^{AB}\, d_s^{A\dagger}\left(N^T\sigma_2\,\sigma_i\tau_2\tau^B\, i\overset{\leftrightarrow}{D}_i N\right)\\ &+ g^{(^3S_1-^3P_1)}\,\epsilon^{ijk}\, d_t^{i\dagger}\left(N^T\sigma_2\,\sigma^k\tau_2\tau_3\overset{\leftrightarrow}{D}^j N\right)\Bigg] + \text{h.c.} + \ldots,\end{aligned} \tag{1}$$

where $a\,\mathcal{O}\overset{\leftrightarrow}{D}_i b = a\,\mathcal{O}D_i b - (D_i a)\mathcal{O}b$, with $\mathcal{O}$ some spin-isospin-operator, and $\mathcal{I} = \text{diag}(1,1,-2)$. The ellipsis stands for higher-order terms, which will not be considered here. The five low-energy constants (LECs) $g^{(X-Y)}$ encode the short-distance details of the interactions. In principle, they can be expressed in terms of parameters of the Standard Model. However, this requires a calculation in the nonperturbative regime of QCD, which has not been feasible yet. As an alternative, the LECs can be obtained by comparison with experimental results. A prerequisite for such a determination is a sufficient number of measured observables in low-energy few-nucleon systems.

3. Two-nucleon systems

Reactions involving two nucleons provide the simplest systems in which to study PV NN interactions. In the case of scattering of a polarized beam and an unpolarized target, the PV longitudinal asymmetry A_L is given by

$$A_L = \frac{\sigma_+ - \sigma_-}{\sigma_+ + \sigma_-}\,, \tag{2}$$

where $\sigma_\pm$ is the total cross section for beams with helicity ± 1. The LO results in EFT($\not\pi$) for various beam and target nucleons are [12, 13, 10]

$$\begin{aligned} A_L^{nn} &= -\sqrt{\frac{32M}{\pi}}\, p\left(g^{(^1S_0-^3P_0)}_{(\Delta I=0)} - g^{(^1S_0-^3P_0)}_{(\Delta I=1)} + g^{(^1S_0-^3P_0)}_{(\Delta I=2)}\right),\\ A_L^{pp} &= -\sqrt{\frac{32M}{\pi}}\, p\left(g^{(^1S_0-^3P_0)}_{(\Delta I=0)} + g^{(^1S_0-^3P_0)}_{(\Delta I=1)} + g^{(^1S_0-^3P_0)}_{(\Delta I=2)}\right),\\ A_L^{np} &= -\sqrt{\frac{32M}{\pi}}\, p\frac{\frac{d\sigma^{^1S_0}}{d\Omega}}{\frac{d\sigma^{^1S_0}}{d\Omega} + 3\frac{d\sigma^{^3S_1}}{d\Omega}}\left(g^{(^1S_0-^3P_0)}_{(\Delta I=0)} - 2g^{(^1S_0-^3P_0)}_{(\Delta I=2)}\right)\\ &\quad - \sqrt{\frac{32M}{\pi}}\, p\,\frac{\frac{d\sigma^{^3S_1}}{d\Omega}}{\frac{d\sigma^{^1S_0}}{d\Omega} + 3\frac{d\sigma^{^3S_1}}{d\Omega}}\left(g^{(^3S_1-^1P_1)} + 2g^{(^3S_1-^3P_1)}\right), \end{aligned} \tag{3}$$

where M is the nucleon mass and p the nucleon momentum in the center-of-mass frame. Coulomb effects are neglected in the case of pp scattering. They can be included consistently in the EFT($\not\pi$) formalism, but as shown in Ref. [10] only amount to corrections on the order of 3% even at the lowest energy for which the asymmetry has been measured experimentally.

Journal of Physics: Conference Series **562** (2014) 012003 doi:10.1088/1742-6596/562/1/012003

The asymmetry in np scattering is related to another PV observable, the spin rotation angle per unit length in the transmission of perpendicularly polarized neutrons through an unpolarized hydrogen target. At next-to-leading order (NLO) in the EFT($\not\pi$) expansion it is given by

$$\frac{1}{\rho}\frac{d\phi^{np}_{\mathrm{PV}}}{dl} = 4\sqrt{2\pi M}\left(\frac{2g^{({}^3S_1-{}^3P_1)} + g^{({}^3S_1-{}^1P_1)}}{\gamma_t}\frac{Z_t+1}{2} + \frac{g^{({}^1S_0-{}^3P_0)}_{(\Delta I=0)} - 2g^{({}^1S_0-{}^3P_0)}_{(\Delta I=2)}}{\gamma_s}\frac{Z_s+1}{2}\right), \tag{4}$$

where ρ is the target density, $\gamma_{t/s}$ are the poles in the NN scattering amplitudes in the ${}^3S_1$ and ${}^1S_0$ channels, respectively, and $Z_{t/s} = \frac{1}{1-\gamma_{t,s}r_{t/s}}$, with $r_{t/s}$ the effective ranges in the corresponding channels.

In addition, reactions including photons can also be used to study hadronic parity violation. The angular asymmetry A_γ of outgoing photons in the radiative capture of polarized neutrons on protons, $\vec{n}p \to d\gamma$, is currently being determined by the NPDGamma collaboration at Oak Ridge National Laboratory [14]. At LO the EFT($\not\pi$) result is given by [15, 11]

$$A_\gamma = \frac{4}{3}\sqrt{\frac{2}{\pi}}\frac{M^{\frac{3}{2}}}{\kappa_1(1-\gamma_t a_s)}g^{({}^3S_1-{}^3P_1)}, \tag{5}$$

where κ_1 is the anomalous nucleon isovector magnetic moment and a_s the scattering length in the ${}^1S_0$ channel.

A complementary and independent observable can be measure in the unpolarized capture of neutrons by measuring the circular polarization P_γ of the outgoing photons, $np \to d\vec{\gamma}$. For exactly reversed kinematics, it is identical to the asymmetry A_L^γ in deuteron break-up by polarized photons, $\vec{\gamma}d \to np$,

$$A_L^\gamma = \frac{\sigma_+ - \sigma_-}{\sigma_+ + \sigma_-}, \tag{6}$$

where $\sigma_\pm$ denotes the total break-up cross section for photons with $\pm$ helicity. The energy dependence of A_L^γ has been determined in EFT($\not\pi$) to NLO in Ref. [16]. A measurement of A_L^γ would place stringent demands on the intensity and control of the beam as well as on the control of systematic effects. The possibility of performing such an experiment at an upgraded High Intensity Gamma-Ray Source (HIGS) at the Triangle Universities Nuclear Laboratory is currently being considered. To aid these efforts, Ref. [16] considers two simplified figures of merit to determine at which energy to best perform a measurement. Using order of magnitude estimates for the unknown PV LECs indicates that the preferred energy range is 2.26 MeV to 2.3 MeV.

4. Three-nucleon systems

In the PC sector of EFT($\not\pi$), naive dimensional analysis does not predict three-nucleon interactions at LO. However, it was shown that a three-nucleon term is enhanced compared to this naive power counting and should be taken into account already at LO [17, 18]. The corresponding LEC can be determined from the experimental value of, e.g., the three-nucleon binding energy or the nd scattering length.

In Ref. [19] it was shown that such an enhancement does not occur in the PV sector of the theory and that PV three-nucleon interactions do not have to be included when considering nucleon-deuteron scattering at LO and NLO. This means that up to an accuracy of approximately 10% pseudoscalar few-nucleon observables can be described in terms of PV two-nucleon interactions alone.

The $\vec{n}d$ spin rotation angle was determined in EFT($\not\pi$) to LO and NLO in Refs. [20, 21], with the NLO results given by [20]

$$\frac{1}{\rho}\frac{d\phi_{\rm PV}^{nd}}{dl} = \Big([16 \pm 1.6]g^{(^3S_1-^1P_1)} + [34 \pm 3.4]g^{(^3S_1-^3P_1)} + [4.6 \pm 1.0](3g^{(^1S_0-^3P_0)}_{(\Delta I=0)} - 2g^{(^1S_0-^3P_0)}_{(\Delta I=1)})\Big)\frac{\rm rad}{\rm MeV^{\frac{1}{2}}}, \tag{7}$$

where the uncertainties represent theoretical error estimates.

5. Conclusions and outlook

Parity-violating NN interactions originate in the interplay of weak and nonperturbative strong quark-quark interactions and therefore provide a unique probe of nonperturbative QCD. At energies below the pion mass, EFT($\not\pi$) can be used as a unified framework to describe PC and PV interactions as well as external currents in few-nucleon reactions. All short-distance physics is encoded in the LECs of the theory. An effort is underway to determine the five LO LECs from a suite of measurements in few-nucleon systems. These numbers provide a nontrivial target that any calculation in the nonperturbative regime has to reproduce. Currently, lattice QCD provides a framework for such calculations, and a first calculation of a PV coupling was performed in Ref. [22]. Further work in this direction will be of great importance.

In addition to the results presented here, there are a number of further observables in four- and five-nucleon systems for which experimental results exists. Examples include $\vec{p}\,^4$He scattering [23] and the neutron spin rotation angle in a ^{4}He target [24]. A calculation in terms of an EFT including dynamical pion degrees of freedom was performed for the PV asymmetry in the charge-exchange reaction $^3\text{He}(\vec{n},p)^3\text{H}$ [25], and the corresponding experiment is planned at Oak Ridge National Laboratory.

Acknowledgments

I am grateful to H. W. Grießhammer, D. R. Phillips, R. P. Springer, and J. Vanasse for their collaboration and many stimulating discussions. This work was supported by the US Department of Energy under grant no. DE-SC0010300.

References

[1] Adelberger E G and Haxton W C 1985 *Ann. Rev. Nucl. Part. Sci.* **35** 501–558
[2] Ramsey-Musolf M J and Page S A 2006 *Ann. Rev. Nucl. Part. Sci.* **56** 1–52 (*Preprint* hep-ph/0601127)
[3] Haxton W and Holstein B 2013 *Prog.Part.Nucl.Phys.* **71** 185–203 (*Preprint* 1303.4132)
[4] Schindler M and Springer R 2013 *Prog.Part.Nucl.Phys.* **72** 1–43 (*Preprint* 1305.4190)
[5] Bowman C, Bowman J and Yuan V 1989 *Phys.Rev.* **C39** 1721–1724
[6] Beane S R, Bedaque P F, Haxton W C, Phillips D R and Savage M J 2000 *In *Shifman, M. (ed.): At the frontier of particle physics, vol. 1** 133–269. (*Preprint* nucl-th/0008064)
[7] Bedaque P F and van Kolck U 2002 *Ann. Rev. Nucl. Part. Sci.* **52** 339–396 (*Preprint* nucl-th/0203055)
[8] Platter L 2009 *Few Body Syst.* **46** 139–171 (*Preprint* 0904.2227)
[9] Girlanda L 2008 *Phys. Rev.* **C77** 067001 (*Preprint* 0804.0772)
[10] Phillips D R, Schindler M R and Springer R P 2009 *Nucl. Phys.* **A822** 1–19 (*Preprint* 0812.2073)
[11] Schindler M R and Springer R P 2010 *Nucl. Phys.* **A846** 51–62 (*Preprint* 0907.5358)
[12] Zhu S L, Maekawa C M, Holstein B R, Ramsey-Musolf M J and van Kolck U 2005 *Nucl.Phys.* **A748** 435–498 (*Preprint* nucl-th/0407087)
[13] Holstein B R 2005 *Fizika.* **B14** 165–216 (*Preprint* nucl-th/0607038)
[14] Gericke M, Alarcon R, Balascuta S, Barron-Palos L, Blessinger C *et al.* 2011 *Phys.Rev.* **C83** 015505
[15] Savage M J 2001 *Nucl. Phys.* **A695** 365–373 (*Preprint* nucl-th/0012043)
[16] Vanasse J and Schindler M R 2014 (*Preprint* 1404.0658)
[17] Bedaque P F, Hammer H W and van Kolck U 1999 *Phys. Rev. Lett.* **82** 463–467 (*Preprint* nucl-th/9809025)
[18] Bedaque P F, Hammer H W and van Kolck U 2000 *Nucl. Phys.* **A676** 357–370 (*Preprint* nucl-th/9906032)

[19] Grießhammer H W and Schindler M R 2010 *Eur. Phys. J.* **A46** 73–83 (*Preprint* 1007.0734)
[20] Grießhammer H W, Schindler M R and Springer R P 2012 *Eur.Phys.J.* **A48** 7 (*Preprint* 1109.5667)
[21] Vanasse J 2012 *Phys.Rev.* **C86** 014001 (*Preprint* 1110.1039)
[22] Wasem J 2012 *Phys.Rev.* **C85** 022501 (*Preprint* 1108.1151)
[23] Lang J, Maier T, Muller R, Nessi-Tedaldi F, Roser T *et al.* 1985 *Phys.Rev.Lett.* **54** 170–173
[24] Snow W, Bass C, Bass T, Crawford B, Gan K *et al.* 2011 *Phys.Rev.* **C83** 022501
[25] Viviani M, Baroni A, Girlanda L, Kievsky A, Marcucci L *et al.* 2014 (*Preprint* 1403.2267)

Investigation of the $D^*_{s2}(2573)^+D^+K^0$ vertex via QCD sum rules

Y. Sarac[1], K. Azizi[2], H. Sundu[3]

[1]Electrical and Electronics Engineering Department, Atilim University, 06836 Ankara, Turkey
[2]Physics Department, Doğuş University, Acıbadem-Kadıköy, 34722 Istanbul, Turkey
[3]Department of Physics, Kocaeli University, 41380 Izmit, Turkey

E-mail: yasemin.sarac@atilim.edu.tr

Abstract. In this work the $D^*_{s2}(2573)^+D^+K^0$ vertex is studied and the coupling constant corresponding to the $D^*_{s2}(2573)^+ \rightarrow D^+K^0$ transition is calculated. The calculation is performed using three point QCD sum rules method and the value of the coupling constant is obtained as $g_{D^*_{s2}DK} = (12.85 \pm 3.85)$ GeV^{-1}. The coupling constant is also used to calculate the decay width and the branching ratio of the considered transition.

1. Introduction

The orbitally excited charmed meson was firstly observed in 1986 [1] and the following past few decades have been an era of the observation of them [2, 3, 4, 5, 6, 7, 8, 9, 10, 11]. During these period there also have been several theoretical studies on the masses, strong and electromagnetic transitions of these mesons (See for instance the Refs. [12, 13, 14, 15] and the references there in). Although $D^*_{s2}(2573)$ meson's quantum numbers are not exactly known the width and decay modes of the $D^*_{s2}(2573)$ state favors $I(J^P) = 0(2^+)$ quantum numbers. Therefore in this work we consider it as a charmed strange tensor meson.

If one compares with the other types of mesons, in literature there are little theoretical studies on the properties of the tensor mesons. In order to gain useful information about the nature and internal structure of tensor mesons, the studies on the properties of these mesons may be helpful. This type of studies may provide a way to test the assumptions of some theoretical calculations and understand the experimental results. In the decays of B meson, the charmed tensor mesons appear as an intermediate state, therefore this type of work may also be helpful to gain information about the B meson. The possibility for a search on the decay properties of D^*_{s2} meson provides us with another motivation to study the properties of these mesons.

This work presents the analysis of the $D^*_{s2}(2573)^+D^+K^0$ vertex. The coupling form factor $g_{D^*_{s2}DK}$ for the considered vertex is calculated using QCD sum rules method [16]. In section 2, we present the details of the calculation made for coupling form factor. Section 3 presents numerical analysis of the coupling constant and the decay with of considered decay.

2. QCD sum rules for the strong coupling form factor

In order to obtain the coupling form factor $g_{D^*_{s2}DK}$ we use the following three-point correlation function:

$$\Pi_{\mu\nu}(p,p',q) = i^2 \int d^4x \int d^4y \; e^{-ip\cdot x} \; e^{ip'\cdot y} \; \langle 0|\mathcal{T}\left(J^D(y)\; J^K(0)\; J^{D^{*\dagger}_{s2}}_{\mu\nu}(x)\right)|0\rangle, \quad (1)$$

were $\mathcal{T}$ is the time ordering operator. The J^D, J^K and $J^{D^*_{s2}}_{\mu\nu}$ are the interpolating fields and can be written in terms of the quark field operators as: $J^D(y) = i\bar{d}(y)\gamma_5 c(y)$, $J^K(0) = i\bar{s}(0)\gamma_5 d(0)$ and $J^{D^*_{s2}}_{\mu\nu}(x) = \frac{i}{2}\left[\bar{s}(x)\gamma_\mu \overleftrightarrow{\mathcal{D}}_\nu (x)c(x) + \bar{s}(x)\gamma_\nu \overleftrightarrow{\mathcal{D}}_\mu (x)c(x)\right]$ where $\overleftrightarrow{\mathcal{D}}_\mu (x)$ is two-side covariant derivative.

The correlation function given in Eq. (1) can be calculated following two different ways. It is calculated in terms of hadronic parameters and this side is called as the physical side of the calculation. And also it is calculated in terms of quark and gluon degrees of freedom by the help of the operator product expansion (OPE) in deep Euclidean region which is the theoretical or QCD side. The match of coefficients of same structure obtained from both sides leads us to the QCD sum rules for the intended physical quantity. To supress the contribution of the higher states and continuum double Borel transformation with respect to the variables p^2 and p'^2 is applied.

To get the physical side one inserts complete sets of appropriate hadronic states into the correlation function with the same quantum numbers as the interpolating currents and obtains:

$$\Pi^{had}_{\mu\nu} = \frac{\langle 0 \mid J^K \mid K(q)\rangle\langle 0 \mid J^D \mid D(p')\rangle\langle D^*_{s2}(p,\epsilon) \mid J^{D^*_{s2}}_{\mu\nu} \mid 0\rangle}{(p^2 - m^2_{D^*_{s2}})(p'^2 - m^2_D)(q^2 - m^2_K)}\langle K(q)D(p') \mid D^*_{s2}(p,\epsilon)\rangle + \cdots , \quad (2)$$

By the usage of the explicit expressions of matrix elements appearing in Eq. (2) which can be parameterized in terms of the leptonic decay constants and strong coupling constant, and after a straight forward algebra one obtains the final form of the correlation function after double Borel transformation (for details see Ref. [17]):

In OPE side, we substitute the explicit forms of the interpolating currents into the correlation function Eq. (1) and after contracting out all quark pairs via Wick's theorem, we get

$$\Pi^{OPE}_{\mu\nu} = \frac{i^5}{2} \int d^4x \int d^4y e^{-ip\cdot x} e^{ip'\cdot y} \left\{ Tr\left[\gamma_5 \; S^{ji}_d(-y)\gamma_5 S^{i\ell}_c(y-x)\gamma_\mu \overleftrightarrow{\mathcal{D}}_\nu (x) S^{\ell j}_s(x)\right] + [\mu \leftrightarrow \nu] \right\}, \quad (3)$$

where $S^{i\ell}_c(x)$, $S^{\ell j}_s(x)$ and $S^{ji}_d(x)$ are the heavy and light quark propagators. After some straight forward calculations (for details see Ref. [17]) for this side we obtain the results in the form;

$$\Pi^{OPE}_{\mu\nu}(p,p',q) = \Pi_1(q^2)p_\mu p_\nu + \Pi_2(q^2)p_\nu p'_\mu + \Pi_3(q^2)p_\mu p'_\nu + \Pi_4(q^2)p'_\mu p'_\nu + \Pi_5(q^2)g_{\mu\nu}, \quad (4)$$

with same dirac structures as in physical side. Since the results are lengthy they will not be given here explicitely.

Equating the coefficients of the same Dirac structure obtained from both sides of the correlation function which is $p_\mu p'_\nu$ in our case, one gets the following sum rules for the coupling form factor

$$g_{D^*_{s2}DK} = \frac{e^{\frac{m^2_{D^*_{s2}}}{M^2}} e^{\frac{m^2_D}{M'^2}} 6(m_c + m_d)(m_d + m_s)(m^2_K - q^2) m_{D^*_{s2}}}{f_{D^*_{s2}} f_D f_K m^2_D m^2_K \left[m^4_D + m^2_D(4m^2_{D^*_{s2}} - 2q^2) + (m^2_{D^*_{s2}} - q^2)^2\right]}$$
$$\times \left\{ \int_{(m_c+m_s)^2}^{s_0} ds \int_{(m_c+m_d)^2}^{s'_0} ds' e^{-\frac{s}{M^2}} e^{-\frac{s'}{M'^2}} \rho^{pert}_3(s,s',q^2) + \widehat{\mathbf{B}}\Pi^{nonpert}_3(q^2) \right\}, \quad (5)$$

where s_0 and s'_0 are continuum thresholds, M^2 and M'^2 are the Borel mass parameters and $\rho_3(s,s',q^2) = \frac{1}{\pi} Im[\Pi_3]$.

3. Numerical Results

For the numerical analysis we used the following input parameters $m_c = (1.275\pm0.025)$ GeV [18], $m_d = (4.8^{+0.5}_{-0.3})$ MeV [18], $m_s = (95 \pm 5\ )$MeV [18], $m_{D^*_{s2}(2573)} = (2571.9 \pm 0.8)$ MeV [18], $m_D = (1869.62\pm0.15)$ MeV [18], $m_K = (493.677\pm0.016)$ MeV [18], $f_D = (206.7\pm8.9)$ MeV [18], $f_K = (156.1 \pm 0.2 \pm 0.8 \pm 0.2)$ MeV [18], $f_{D^*_{s2}(2573)} = (0.0230 \pm 0.0011)$ [14], $\langle \frac{\alpha_s G^2}{\pi} \rangle = (0.012 \pm 0.004)$ GeV^4.

Borel mass parameters M^2 and M'^2 and continuum thresholds s_0 and s'_0 appearing in our results are auxiliary parameters, which are not physical parameters. Therefore we need to determine the working regions of them from our analysis seeking weak dependency of our results on them. From our analysis we obtained their working regions as 8.5 GeV$^2 \leq s_0 \leq 9.4$ GeV^2, 4.7 GeV$^2 \leq s'_0 \leq 5.6$ GeV^2, 3 GeV$^2 \leq M^2 \leq 8$ GeV^2 and 2 GeV$^2 \leq M'^2 \leq 5$ GeV^2.

The coupling constant is defined as the value of the form factor at $Q^2 = -q^2 = -m_K^2$ which is outside of the reliable region of our sum rules calculations. Therefore to extend our result we use the fit function $g_{D^*_{s2}DK}(Q^2) = c_1 \exp\Big[-\frac{Q^2}{c_2}\Big] + c_3$ where $Q^2 = -q^2$ and obtain the fit parameters, $c_1 = (12.03 \pm 3.61)$ GeV^{-1}, $c_2 = (12.73 \pm 3.18)$ GeV^2 and $c_3 = (0.81 \pm 0.24)$ GeV^{-1} from our analysis. And from the fit function we obtain the coupling constant as $g_{D^*_{s2}DK}(Q^2 = -m_K^2) = (12.85 \pm 3.85)$ GeV^{-1}.

After determination of the coupling constant the final task of our work is to obtain the decay width and the branching ratio of the considered transition using the value of the coupling constant and the decay with formula given in Ref. [17]. The decay width is achieved as $\Gamma = (1.84 \pm 0.48) \times 10^{-3}$ GeV and using the total width of the initial particle, $\Gamma_{D^*_{s2}(2573)^0} = (17 \pm 4)$ MeV [18], the branching ratio is obtained as $BR = (1.08 \pm 0.27) \times 10^{-1}$.

Acknowledgement

This work has been supported in part by the Scientific and Technological Research Council of Turkey (TUBITAK) under the research project 114F018.

References

[1] Albrecht H et al. (ARGUS Collaboration) 1986 *Phys. Rev. Lett.* **56** 549.
[2] H. Albrecht et al. (ARGUS Collaboration) 1989 *Phys. Lett. B* **232** 398; Albrecht H et al. (ARGUS Collaboration) 1989 *Phys. Lett. B* **231** 208; Albrecht H et al. (ARGUS Collaboration) 1989 *Phys. Lett. B* **221** 422; Albrecht H et al. (ARGUS Collaboration) 1989 *Phys. Lett. B* **230** 162; Albrecht H et al. (ARGUS Collaboration) 1992 *Phys. Lett. B* **297** 425.
[3] Frabetti P L et al. (E687 Collaboration) 1994 *Phys. Rev. Lett.* **72** 324.
[4] Anjos J C et al. (E691 Collaboration) 1989 *Phys. Rev. Lett.* **62** 1717.
[5] Alexander J P et al (CLEO Collaboration) 1993 *Phys. Lett. B* **303** 377.
[6] Avery P et al. (CLEO Collaboration) 1990 *Phys. Rev. D* **41** 774; Avery P et al. (CLEO Collaboration) 1994 *Phys. Lett. B* **331** 236.
[7] Kubota Y et al. (CLEO Collaboration) 1994 *Phys. Rev. Lett.* **72** 1972.
[8] Bergfeld T et al. (CLEO Collaboration) 1994 *Phys.Lett. B* **340** 194.
[9] Link J et al. (FOCUS Collaboration) 2004 *Phys. Lett. B* **586** 11.
[10] Abe K et al. (BELLE Collaboration) 2004 *Phys. Rev. D* **69** 112002.
[11] Abazov V M et al. (D0 Collaboration) 2005 *Phys. Rev. Lett.* **95** 161602.
[12] Godfrey S 2005 *Phys. Rev. D* **72** 054029.
[13] Sundu H and Azizi K 2012 *Eur. Phys. J. A* **48** 81.
[14] Azizi K, Sundu H, Süngü J Y and Yinelek N 2013 *Phys. Rev. D* **88** 036005 ; 2013 *Phys. Rev. D* **88**, 099901(E).
[15] Azizi K, Sundu H, Türkan A Y and Veli Veliev E 2014 *J.Phys. G* **41** 035003.
[16] Shifman M A, Vainshtein A I and Zakharov V I 1979 *Nucl. Phys. B* **147** 385 ; 1979 *Nucl. Phys. B* **147** 448.
[17] Azizi K, Sarac Y and Sundu H 2014 arXiv:1402.6887.
[18] Beringer et al. (Particle Data Group) 2012 *Phys. Rev. D* **86** 010001 and 2013 partial update for the 2014 edition.

4th International Hadron Physics Conference (TROIA'14) IOP Publishing
Journal of Physics: Conference Series **562** (2014) 012005 doi:10.1088/1742-6596/562/1/012005

Relations between the QCD sum rules for BBV couplings and vector-dominance model

Valeri Zamiralov
Skobeltsyn Institute of Nuclear Physics, Lomonosov Moscow State University Moscow,Russia
E-mail: zamir@depni.sinp.msu.ru

Abstract. Story of QCD sum rules is briefly reviewed. The dispersion superconvergence sum rule is shown to relate πN coupling constant with the 33-resonance width without unitary symmetry arguments. The QCD sum rules instead reveal the unitary symmetry pattern as all the correlation functions can be reduced to 2 ones repeating the F and D couplings for the octet baryons $1/2^+$. Strong relations between the correlators are established. Vector dominance hypothesis allows to put new restrictions to the QCD sum rules for VBB couplings.

1. Introduction

The QCD sume rules begin their story from the dispersion sum rules formulated more than half century ago. I would remind superconvergent sum rules (SCSR) which have been formulated in 1966 [1] for elastic πN scattering. Assuming that spin-flip amplitude B goes to zero faster than $1/E$ the Cauchy integral over the closed contour with the cuts would be zero. Upon disregarding the integral over the large circle one get

$$\int_{-\infty}^{\infty} Im\, B(E,\theta)dE = 0.$$

With saturation of the integral with the nucleon pole and 33-resonance contribution [1] πN coupling constant was related to the width of the 33-resonance resulting in $\Gamma_{\Delta N\pi}$=117MeV with $g^2_{\pi N}/4\pi = 14.6$ in nice agreement with the $\Gamma^{exp}_{\Delta N\pi} = 116-118$ MeV. It was the beginning. The next important step was the construction of the FESR - Finite Energy Sum Rules [2]. Let A be some amplitude and A_R be its high-energy model behavior and let it be, say, Regge - model one: $A_R = \beta(t)s^{\alpha(t)}$. For their difference one can construct SCSR:

$$\int (A(s,t) - A_R(s,t))ds = 0. \tag{1}$$

If for $s \leq R$ the A and A_R become practically equal one arrives at the FESR;

$$\int_{thresh}^{R} A(s.t)ds = \frac{\beta(t)}{\alpha(t)+1}R^{\alpha(t)+1}. \tag{2}$$

The LHS can be calculated through some phenomenological quantities while the RHS can be calculated analytically in terms of high-energy model parameters.

2. QCD sum rules

With arrival of QCD as the theory of strong interaction basing on quark and gluons one soon has undersood that in the region far from the perturbative expansion it is impossible to advance without some kind of non-perturbative theory. The way was found in a similar fashion to the FESR's: Shifman, Vainstein and Zakharov in 1979 [3] have proposed the sum rule where the LHS could be calculated by inserting phenomenological states while the RHS could be rather safely calculated in terms of well-defined operators of definite dimensions.

In order to maintain only the ground states and to suppress the high-dimensional contributions they used Borel transformation and arrived at the well-controlled sum rules with a small number of universal parameters, namely, non-perturbative vacuum expectation values (vev's) of the operators of definite dimensions like $a_q = -(2\pi)^2\langle 0|\bar{q}q|0\rangle$, $a_q m_0^2 = (2\pi)^2 g_s\langle 0|\bar{q}\sigma \cdot Gq|0\rangle$, $b = (2\pi)\langle 0|GG|0\rangle$, as well as characteristic QCD parameter $\Lambda_{QCD} \approx 100$ MeV and normalization point $\mu \approx 0.5$ GeV to which the used values of condensates are referred.

There are mainly two free parameters left that is the so called Borel parameter M^2 and the threshold parameter s_0 which effectively controls the suppression of the continuous spectrum and high-excited states. One more parameter was introduced connected with a possibility to choose various so-called interpolating currents.

The basic notion in QCD sum rules is the polarization operator or correlator function [3], [4]. The correlator for baryon-baryon-vector meson vertex would have the form

$$\Pi(p) = \int e^{ipx}\langle 0|T(\eta(x)\bar{\eta}(0))|V\rangle dx, \tag{3}$$

where η is the so-called correlation current and should be constructed in QCD in terms of quark fields. The Σ^0 interpolating current can be written :

$$\eta^{\Sigma^0} = \sqrt{\frac{1}{2}}\epsilon^{abc}\Big[\Big(u^{aT}Cs^b\Big)\gamma_5 d^c - \Big(s^{aT}Cd^b\Big)\gamma_5 u^c + \beta\Big(u^{aT}C\gamma_5 s^b\Big)d^c - \beta\Big(s^{aT}C\gamma_5 d^b\Big)u^c\Big], \tag{4}$$

where C is the charge conjugation operator, a, b, c, d, e, f are the color indices and β is an arbitrary parameter, $\beta = -1$ corresponds to the Ioffe current [4]. The currents of the other baryons can be obtained from Σ^0 current by making appropriate substitutions of quarks. The Λ interpolating current can be obtained through $\sqrt{3}\eta^{\Lambda} = -\eta^{\Sigma^0} + 2\eta^{\Sigma^0(d\leftrightarrow s)}$.

3. Quark-biquark model and SU(3)

To obtain unitary symmetry results in the framework of quark scheme we invoke quark-biquark model that is assume that photon (or any other boson quantum) distinguishes between the (equal) quarks which form a biquark and a single quark [5]. The final formula for the magnetic moment of any hyperon with the Σ-like quark wave function would have the form

$$\mu(\Sigma(ud, Q)) = (e_u + e_d)\mu_F + e_Q(\mu_F - \mu_D), \tag{5}$$

where μ_F and μ_D are usual SU(3) quantities. Important result of this model is that formal group-theoretical couplings of the F and D type have in it a well-defined physical meaning.

In a similar way new formulae for the meson-baryon coupling constants can be written in terms of the quark couplings with mesons and SU(3) constants F and D. A coupling constant of any Σ-like hyperon would have the form

$$G(\Sigma(ud, Q)) = (g_{Muu} + g_{Mdd})F + g_{MQQ}(F - D). \tag{6}$$

It is a little astonishing that these formulas are valid also for the QCD sum rules.

4. Relations between Σ and Λ quantities in quark model and QCD

To establish unitary symmetry relations between Sigma and Lambda quantities we remind that isospin wave functions $|\Sigma^0\rangle$ and $|\Lambda\rangle$ are related to the ones of U- and V- spins as

$$-2|\Lambda_{us}\rangle = \sqrt{3}|\Sigma^0\rangle + |\Lambda\rangle, \qquad -2|\Lambda_{ds\rangle} = -\sqrt{3}|\Sigma^0\rangle + |\Lambda\rangle, \tag{7}$$

$$2|\Sigma_{us}\rangle = -|\Sigma^0\rangle + \sqrt{3}|\Lambda\rangle, \qquad 2|\Sigma_{ds}\rangle = |\Sigma^0\rangle + \sqrt{3}|\Lambda\rangle.$$

where ds (us) means interchange of flavores $d(u)$ and s.

For any operator $\mathcal{O}$ the following relations are valid between the matrix elements:

$$2\langle\Sigma_{us}|\mathcal{O}|\Sigma_{us}\rangle + 2\langle\Sigma_{us}|\mathcal{O}|\Sigma_{us}\rangle - \langle\Sigma^0|\mathcal{O}|\Sigma^0\rangle = 3\langle\Lambda|\mathcal{O}|\Lambda\rangle, \tag{8}$$

$$-2\langle\Sigma_{us}|\mathcal{O}|\Sigma_{us}\rangle + 2\langle\Sigma_{us}|\mathcal{O}|\Sigma_{us}\rangle = \sqrt{3}(\langle\Sigma^0|\mathcal{O}|\Lambda\rangle + \langle\Lambda|\mathcal{O}|\Sigma^0\rangle),$$

and similar ones with interchange of the Λ and Σ symbols. We introduce also a new relation which reads as

$$\langle\Sigma\Lambda\rangle|_{us} + \langle\Sigma\Lambda\rangle|_{ds} = \langle\Sigma\Lambda\rangle.$$

As we have noted already these relations are simple but in very tedious QCD sum rule calculations they turn to be very useful. We write now the analogous relations for the QCD correlators [6]:

$$2[\tilde{\Pi}^{\Sigma^0(ds)} + \tilde{\Pi}^{\Sigma^0(us)}] - \Pi^{\Sigma^0} = 3\Pi^{\Lambda}, \qquad 2[\tilde{\Pi}^{\Lambda(ds)} + \tilde{\Pi}^{\Lambda(us)}] - \Pi^{\Lambda} = 3\Pi^{\Sigma^0}. \tag{9}$$

$$[\tilde{\Pi}^{\Sigma^0(ds)} - \tilde{\Pi}^{\Sigma^0(us)}] = \sqrt{3}\Pi^{\Lambda\Sigma^0}, \qquad \tilde{\Pi}^{\Lambda(ds)} - \Pi^{\tilde{\Lambda}(us)} = -\sqrt{3}\Pi^{\Lambda\Sigma^0}. \tag{10}$$

Our new relation for correlation functions reads

$$\Pi^{\Lambda\Sigma^0(us)} + \Pi^{\Lambda\Sigma^0(ds)} = \Pi^{\Lambda\Sigma^0} \tag{11}$$

5. QCD sum rules for the vector meson–baryon coupling constants

The following correlation function is considered [7] (see also,e.g., [8],[9]):

$$\Pi^{B_1\to B_2 V} = i\int d^4x e^{ipx}\,\langle V(q)\,|\mathcal{T}\,\{\eta_{B_2}(x)\bar{\eta}_{B_1}(0)\}|\,0\rangle \tag{12}$$

The correlator can be calculated in terms of the hadrons, as well as in the deep Euclidean region $p^2 \to -\infty$, in terms of the quark and gluon degrees of freedom.

To construct the phenomenological part of the correlator we insert a complete set of intermediate states with the quantum numbers of current operators η_B.

After isolating the ground state baryons we have:

$$\Pi^{B_1\to B_2 V}(p_1^2, p_2^2) = \frac{\langle 0\,|\eta_{B_2}|\,B_2(p_2)\rangle}{p_2^2 - m_2^2}\,\langle B_2(p_2)V(q)\,|\,B_1(p_1)\rangle\,\frac{\langle B_1(p_1)\,|\bar{\eta}_{B_1}|\,0\rangle}{p_1^2 - m_1^2} + \cdots\,, \tag{13}$$

$p_1 = p_2 + q$, m_i is the mass of baryon B_i, and $\cdots$ means contributions of the higher states and the continuum. Matrix elements are defined as [7]:

$$\langle 0\,|\eta_{B_i}|\,B_i(p_i)\rangle = \lambda_{B_i} u(p_i)\;, \langle B_2(p_2)V(q)\,|\,B_1(p_1)\rangle = \bar{u}(p_2)\left[f_1\gamma_\mu - f_2\frac{i}{m_1+m_2}\sigma_{\mu\nu}q^\nu\right]u(p_1)\varepsilon^\mu\;, \quad (\text{:}$$

λ_{B_i} is the overlap amplitude for baryon B_i [4]. We choose in (13) structures $\not p \not\varepsilon \not q$ and $\not p(\varepsilon \cdot p)$ as they exhibit better convergence. QCD sum rules for vector-meson baryon coupling constants for electric- and magnetic- type couplings are then [7]

$$f_1 = \frac{\kappa}{\lambda_B^2}\Pi^{f_1}(u,d,s;M^2,s_0,\beta) \qquad f_{1+2} = f_1 + f_2 = \frac{\kappa}{\lambda_B^2}\Pi^{f_1+f_2} \quad \kappa = e^{-(m_B^2/M^2)-(m_V^2/M^2)}. \tag{15}$$

Correlators for VBB coupling constants can be written in terms of only two invariant functions for each coupling, of the electric and magnetic type. For $\Sigma^0\Sigma^0 V$ it reads similar to (6)

$$\Pi^{\Sigma^0\Sigma^0 V}(u,d,s;M^2,s_0,\beta) = g_{Vuu}\Pi_1(u,d,s;M^2,s_0,\beta) \tag{16}$$

$$+g_{Vdd}\Pi_1(d,u,s;M^2,s_0,\beta) + g_{Vss}\Pi_2(u,d,s;M^2,s_0,\beta).$$

6. Unitary symmetry and VDM

Now we try to relate baryon charges, magnetic moments and strong coupling constants with light vector mesons. We start from the Lagrangian for VBB couplings maintaining only unitary indices but taking in mind that there are two kinds of VBB couplings, electric $g_{VBB}^{(e)}$ and magnetic $g_{VBB}^{(m)}$ ones.

$$\sqrt{2}\mathcal{L}^{VBB} = (F+D)\bar{B}^\alpha_\beta B^\gamma_\alpha V^\beta_\gamma + (D-F)\bar{B}^\beta_\alpha B^\alpha_\gamma V^\gamma_\beta - (D-F)\cdot Tr(\bar{B}B)\cdot TrV\,, \tag{17}$$

that is $g(\rho^0 pp) = F + D$, $g(\omega pp) = 3F - D$, $g(\phi pp) = 0$, etc. We write the Lagrangian for the direct $V\gamma$ interaction as

$$\mathcal{L}^{V\gamma} = \frac{\sqrt{2}}{g_{V\gamma}}(V_1^1 - \frac{1}{3}TrV) = \frac{1}{g_{V\gamma}}(\rho^0 + \frac{1}{3}\omega - \frac{\sqrt{2}}{3}\phi), \tag{18}$$

that is $g_{\rho^0\gamma} = g_{V\gamma}$, $g_{\omega\gamma} = 3g_{\rho^0\gamma}$, $g_{\phi\gamma} = -3g_{\rho^0 V\gamma}/\sqrt{2}$. Now we construct baryon electric charges and magnetic moments using formula

$$e_B, \mu_B = \frac{g_{\rho BB}^{(e,m)}}{g_{\rho\gamma}} + \frac{g_{\omega BB}^{(e,m)}}{g_{\omega\gamma}} + \frac{g_{\phi BB}^{(e,m)}}{g_{\phi\gamma}}. \tag{19}$$

We write it for magnetic moments for the proton using Eq.(18):

$$\mu_p = \frac{F^m + D^m}{g_{\rho\gamma}} + \frac{3F^m - D^m}{g_{\omega\gamma}} = \frac{2}{g_{\rho\gamma}}(F^m + \frac{1}{3}D^m) = \mu_F + \frac{1}{3}\mu_D. \tag{20}$$

Baryon octet magnetic moments are given by $SU(3)$ with renormalized F and D couplings $\mu_F = 2F^m/g_{\rho\gamma}$, $\mu_D = 2D^m/g_{\rho\gamma}$. Electric charges of baryons are obtained by changing μ_F to e and putting μ_D=0 and /or deduced from the formulas in terms of D^e and F^e similar with D^e=0 and $2F^e$=$g_{\rho\gamma}$. This equality gives us criterium to distinguish between the models if we believe in Vector Dominance Model (VDM). We proceed with the work [10]. Sum rules for the "electric" structure $p_\mu\gamma_\nu + p_\nu\gamma_\mu$ Eqs.(24-27) of [10] for the Σ^0 and Λ in a little simplified form read

$$(g_{\Sigma^0}^V + C_\Sigma)\tilde{\lambda}_\Sigma^2 e^{-m_\Sigma^2/M^2} = M^6 E_1^\Sigma L^{-4/9}(g_u + g_d + g_s) + \frac{4}{3}L^{4/9}[(g_u + g_d)a_u a_d + g_s a_u a_d] + \tag{21}$$

$$M^2[-2m_s a_s(g_u + g_d) - 4m_u a_u g_d - 4m_d a_d g_u + 4m_u a_s g_d + 4m_d a_s g_u];$$

$$(g_\Lambda^V + C_\Lambda)\tilde{\lambda}_\Lambda^2 e^{-m_\Lambda^2/M^2} = M^6 E_1^\Lambda L^{-4/9}(g_u + g_d + g_s) + \frac{4}{9}L^{4/9}[-(g_u + g_d)a_u a_d + 2(g_u a_d + g_d a_u)a_s + \tag{22}$$

$$g_s(a_u + a_d)a_s - g_s a_u a_d] - 2m_s a_s(g_u + g_d) + \frac{8}{3}m_s(a_u g_d + a_d g_u)$$

What one sees in this formula of particular interest? (We omit the contribution of the vev ζ of [10] due to symmetrical Lorentz indices $\mu\nu$.) Disregarding for simplicity quark masses one obtains the sum rules with overall factor $(g_u + g_d + g_s)$. Generally speaking this is not the case for many other works on the subject even with zero quark masses. That is very often for the "electric"-like coupling the "electric" charge does not factored out in the overall formula. It looks strange and we are trying now to study this situation for VBB couplings more attentively.

Now we apply our reasoning about VDM and obtain the electric charges of the octet baryons. The result with the use of the experimental values of $V\gamma$ constants $g_{\rho\gamma} = 5.05$, $g_{\omega\gamma} = 17.02$, $g_{\phi\gamma} = 12.89$ was not very encouraging. But if we take now another set of $V\gamma$ constants and shift $g_{\phi\gamma}^{exp}$=12.89 to the value 9.9, $g_{\rho\gamma}^{exp}$=5.05 to 4.8 and $g_{\omega\gamma}^{exp}$=17.02 to 14.4 the VDM fit with a "click" became exact and we reproduce the charge pattern of the baryons through ρ, ω and ϕ contributions (see Eq.(43) and Table 1 in [10]):

$$e(p) = \frac{2.4^\rho}{4.8} + \frac{7.2^\omega}{14.4} = 1; \qquad e(n) = -\frac{2.4^\rho}{4.8} + \frac{7.2^\omega}{14.4} = 0; \tag{23}$$

$$e(\Sigma^\pm) = \pm\frac{4.8^\rho}{4.8} + \frac{4.8^\omega}{14.4} - \frac{3.3^\phi}{9.9} = \pm 1; \quad e(\Sigma^0) = -e(\Lambda) = \frac{4.8^\omega}{14.4} - \frac{3.3^\phi}{9.9} = 0$$

$$e(\Xi^0) = \frac{2.4^\rho}{4.8} + \frac{2.4^\omega}{14.4} - \frac{6.6^\phi}{9.9} = 0; \qquad e(\Xi^-) = -\frac{2.4^\rho}{4.8} + \frac{2.4^\omega}{14.4} - \frac{6.6^\phi}{9.9} = -1.$$

We can deduce from these results that (1) - the QCD sum rules of [10] are very truthful, (2) - the values of the constants VBB seem to be away of the real values by amounts of the discrepancy between the experimental values of $g_{V\gamma}$ and the effective values used in our relations. These calculations show in a particularly spectacular way the inner consistency of the QCD results of [10] and prove our idea that with VDM hypothesis it is possible to make conclusions as to the reliability of QCD sum rule results.

7. Conclusion

The QCD sum rules for the baryon coupling with vector mesons reveal the unitary symmetry pattern as all the correlation functions can be reduced to 2 ones in full analogy with F and D couplings for the octet baryons $1/2^+$. Relations between the correlators are established. Vector dominance hypothesis allows to put new restrictions on the QCD sum rules for VBB couplings.

Acknowledgments

The author would like to thank Troia'14 organizers for hospitality and partial support.

[1] Soloviev L D 1966 *Sov. Nucl. Phys.* **2** 598
[2] Logunov A A, Soloviev L D, Tavkhelidze A N 1967 *Phys. Lett.* B**24** 181
[3] Shifman M A, Vainshtein V I and Zakharov V I 1979 *Nucl. Phys.* **B147** 385
[4] Ioffe B L *Nucl. Phys.* 1981 B**188** 317
[5] Gelmi L, Lepshokov S N and Zamiralov V S, 1989 *Moscow University Bulletin, Physics. Astronomy* **2** 33
[6] Ozpineci A, Yakovlev S B and Zamiralov V S 2005 *At.Nucl.Phys.* **68** 304
[7] Aliev T M, Ozpineci A, Savci M and Zamiralov V S 2009 *Phys. Rev.* **D80** 016010
[8] Lai Wang and Lee F X 2008 *Phys. Rev.* **D78** 013003
[9] Shi-lin Zhu, Hwang W Y, and Yang Z 1998 *Phys. Rev.* **D57** 1524
[10] Erkol G, Timmermans R G E and Rijken Th A 2006 *Phys. Rev.* **C74** 045201

Journal of Physics: Conference Series **562** (2014) 012006 doi:10.1088/1742-6596/562/1/012006

The double radiative $\bar{B} \to X_s\gamma\gamma$ decay at $O(\alpha_s)$ in QCD

Ahmet Kokulu

Department of Mathematical Sciences, University of Liverpool, L69 3BX Liverpool, United Kingdom

E-mail: akokulu@liverpool.ac.uk

Abstract. In these proceedings, we briefly review the individual interference contribution of the electromagnetic dipole operator $\mathcal{O}_7$ to the double differential decay width $d\Gamma_{77}/(ds_1\, ds_2)$ for the process $\bar{B} \to X_s\gamma\gamma$ at $O(\alpha_s)$ in QCD, which is based on our work in [1]. We define two kinematical variables s_1 and s_2 as $s_i = (p_b - q_i)^2/m_b^2$, where p_b, q_1, q_2 are the momenta of b-quark and two photons. While the (renormalized) virtual corrections are worked out exactly for a certain range of s_1 and s_2, we retained in the gluon bremsstrahlung process only the leading power w.r.t. the (normalized) hadronic mass $s_3 = (p_b - q_1 - q_2)^2/m_b^2$ in the underlying triple differential decay width $d\Gamma_{77}/(ds_1 ds_2 ds_3)$. We found that the double differential decay width, based on this approximation, is free of infrared- and collinear singularities when summing up the virtual- and real-radiation corrections, while this was not the case when keeping all powers in s_3 in the gluon bremsstrahlung process due to the configurations allowing collinear photon emission from the (massless) s-quark. Lastly, we compare our analytical results with those obtained in a recently extended work [2], where a non-zero strange quark mass was introduced to regulate the collinear photon configurations.

1. Introduction

Inclusive rare B-meson decays provide a crucial place to probe new physics indirectly. In the Standard Model (SM) all these processes proceed through loop diagrams (due to Glashow-Iliopoulos-Maiani mechanism) and thus are relatively suppressed. In the extensions of the SM the contributions stemming from the diagrams with "new" particles in the loops can be comparable or even larger than the contribution from the SM. Thus getting experimental information on rare decays puts stringent constraints on the extensions of the SM or can even lead to a disagreement with the SM predictions, providing evidence for some "new physics".

To make a rigorous comparison between experiment and theory, precise SM calculations for the (differential) decay rates are mandatory. While the branching ratios for $\bar{B} \to X_s\gamma$ [3] and $\bar{B} \to X_s\ell^+\ell^-$ are known today even to next-to-next-to-leading logarithmic (NNLL) precision (for reviews, see [4, 5]), other branching ratios, like the one for $\bar{B} \to X_s\gamma\gamma$ discussed in these proceedings, has been calculated before to leading logarithmic (LL) precision in the SM by several groups [6, 7, 8, 9] and only recently a first step towards next-to-leading-logarithmic (NLL) precision was presented by us in [1]. In contrast to $\bar{B} \to X_s\gamma$, the current-current operator $\mathcal{O}_2$ has a non-vanishing matrix element for $b \to s\gamma\gamma$ at order α_s^0 precision, leading to an interesting interference pattern with the contributions associated with the electromagnetic dipole operator $\mathcal{O}_7$ already at LL precision. As a consequence, potential new physics should be clearly visible not only in the total branching ratio, but also in the differential distributions.

As the process $\bar{B} \to X_s\gamma\gamma$ is expected to be measured at the planned Super B-factory in Japan (**SuperKEKB**), it is necessary to calculate the differential distributions to NLL precision in the SM, in order to fully exploit its potential concerning new physics.

While the Wilson coefficients $C_i(\mu)$ (appearing in the definition of the effective Hamiltonian) are known to sufficient precision at the low scale $\mu \sim m_b$ since a long time (see e.g. the reviews [4, 5] and references therein), the matrix elements $\langle s\gamma\gamma|\mathcal{O}_i|b\rangle$ and $\langle s\gamma\gamma\, g|\mathcal{O}_i|b\rangle$, which in a NLL calculation are needed to order g_s^2 and g_s, respectively, are not fully known yet. To calculate the $(\mathcal{O}_i, \mathcal{O}_j)$-interference contributions to the differential distributions at order α_s is in many respects of similar complexity as the calculation of the photon energy spectrum in $\bar{B} \to X_s\gamma$ at order α_s^2 needed for the NNLL computation. As a first step in this NLL enterprise, we derived in our paper [1], the $O(\alpha_s)$ corrections to the $(\mathcal{O}_7, \mathcal{O}_7)$-interference contribution to the double differential decay width $d\Gamma/(ds_1 ds_2)$ at the partonic level. We defined the variables s_1 and s_2 as $s_i = (p_b - q_i)^2/m_b^2$, where p_b and q_i denote the four-momenta of the b-quark and the two photons, respectively.

At order α_s there are contributions to $d\Gamma_{77}/(ds_1 ds_2)$ with three particles (s-quark and two photons) and four particles (s-quark, two photons and a gluon) in the final state. These contributions correspond to specific cuts of the b-quarks self-energy at order $\alpha^2 \times \alpha_s$, involving twice the operator $\mathcal{O}_7$. As there are additional cuts, which contain for example only one photon, our observable cannot be obtained using the optical theorem, i.e., by taking the absorptive part of the b-quark self-energy at three-loop. We therefore calculated the mentioned contributions with three and four particles in the final state individually.

We work out the QCD corrections to the double differential decay width in the kinematical range (see the left frame of Fig. 1)

$$0 < s_1 < 1 \quad ; \quad 0 < s_2 < 1 - s_1 \, .$$

Concerning the virtual corrections, all singularities (after ultra-violet renormalization) are due to **soft gluon** exchanges and/or **collinear gluon** exchanges involving the s-quark. Concerning the bremsstrahlung corrections (restricted to the same range of s_1 and s_2), there are in addition kinematical situations where **collinear photons** are emitted from the s-quark. The corresponding singularities did not cancel when combined with the virtual corrections. We found, however, that there are no singularities associated with collinear photon emission in the double differential decay width when only retaining the leading power w.r.t to the (normalized) hadronic mass $s_3 = (p_b - q_1 - q_2)^2/m_b^2$ in the underlying triple differential distribution $d\Gamma_{77}/(ds_1 ds_2 ds_3)$. The results of our paper [1] were obtained within this "approximation".

2. The leading and the $O(\alpha_s)$ results for the decay width

In $d = 4$ dimensions, the leading-order spectrum (in our restricted phase-space) is given by

$$\frac{d\Gamma_{77}^{(0)}}{ds_1\, ds_2} = \frac{\alpha^2\, \bar{m}_b^2(\mu)\, m_b^3\, |C_{7,eff}(\mu)|^2\, G_F^2\, |V_{tb}V_{ts}^*|^2\, Q_d^2}{1024\,\pi^5} \frac{(1 - s_1 - s_2)}{(1 - s_1)^2 s_1 (1 - s_2)^2 s_2}\, r_0 \, . \quad (1)$$

where

$$r_0 = -48 s_2^3 s_1^3 + 96 s_2^2 s_1^3 - 56 s_2 s_1^3 + 8 s_1^3 + 96 s_2^3 s_1^2 - 192 s_2^2 s_1^2 + 112 s_2 s_1^2 - 56 s_2^3 s_1 + 112 s_2^2 s_1 - 96 s_2 s_1 + 8 s_1 + 8 s_2^3 + 8 s_2$$

Adding the renormalized virtual corrections and the bremsstrahlung corrections, the complete order α_s correction to the double differential decay width $d\Gamma_{77}/(ds_1\, ds_2)$ reads

$$\frac{d\Gamma_{77}^{(1)}}{ds_1\, ds_2} = \frac{\alpha^2\, \bar{m}_b^2(\mu)\, m_b^3\, |C_{7,eff}(\mu)|^2\, G_F^2\, |V_{tb}V_{ts}^*|^2\, Q_d^2}{1024\,\pi^5} \times \frac{\alpha_s}{4\pi} C_F \left[\frac{-4\, r_0\, (1 - s_1 - s_2)}{(1 - s_1)^2\, s_1\, (1 - s_2)^2\, s_2} \log \frac{\mu}{m_b} + f \right], \quad (2)$$

where the explicit expression for f can be found in [1].

The order α_s correction $d\Gamma_{77}^{(1)}/(ds_1 ds_2)$ in Eq. (2) to the double differential decay width for $b \to X_s\gamma\gamma$ was the main result of our paper [1].

3. Numerical illustrations I

In our procedure the NLL corrections have three sources: (a) α_s corrections to the Wilson coefficient $C_{7,eff}(\mu)$, (b) expressing $\bar{m}_b(\mu)$ in terms of the pole mass m_b and (c) virtual- and real- order α_s corrections to the matrix elements. To illustrate the effect of source (c), which was worked out for the first time in our paper [1] (see also Ref. [10] for a brief review), we show in the middle frame of Fig. 1 (by the long-dashed line) the (partial) NLL result in which source (c) is switched off. We conclude that the effect (c) is roughly of equal importance as the combined effects of (a) and (b).

For completeness we show in the middle frame of this figure (by the dotted line) also the result when QCD is completely switched off, which amounts to put $\mu = m_W$ in the LL result. Furthermore, the right frame of Fig. 1 illustrates the LL spectrums for the cases when taking into account the ($\mathcal{O}_7$, $\mathcal{O}_7$) interference contribution only (dashed curve) and also when all contributions associated with $\mathcal{O}_1$, $\mathcal{O}_2$ and $\mathcal{O}_7$ are considered (solid curve).

From Fig. 1 we see that the NLL results are substantially smaller (typically by 50% or slightly more) than those at LL precision, which was also the case when choosing other values for s_2.

In the numerical discussion above, we observed the following: Generally speaking, NLL corrections are not small, when taking into account the full range of μ, i.e., $m_b/2 < \mu < 2m_b$.

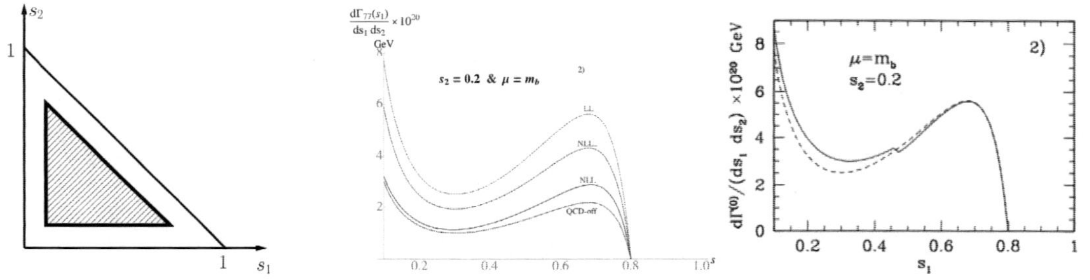

Figure 1. Left: The relevant phase-space region for s_1 and s_2 used in this paper. **Middle:** Double differential decay width, based on the operator $\mathcal{O}_7$ only, as a function of s_1 for s_2 fixed at $s_2 = 0.2$ and $\mu = m_b$. The dotted(black), the short-dashed(red) and the solid line(blue) shows the result when neglecting QCD-effects, the LL and the NLL result, respectively. The long-dashed line(purple) represents the (partial) NLL result in which the virtual- and bremsstrahlung corrections worked out in our paper [1] are switched off. **Right:** Plot taken from Ref. [2] illustrating the LL decay width as a function of s_1 for s_2 fixed at $s_2 = 0.2$ and $\mu = m_b$. In this frame, the dashed line shows the result when only the ($\mathcal{O}_7$, $\mathcal{O}_7$) interference is taken into account, while the solid line shows the result when all contributions associated with $\mathcal{O}_1$, $\mathcal{O}_2$ and $\mathcal{O}_7$ are included.

4. Numerical illustrations II

In this section we briefly compare the result of the present work [1] with those obtained in a recent analysis [2]. In the work of [2], a non-zero mass for the s-quark was introduced in order to regulate the collinear photon emissions from the s-quark, retaining all powers w.r.t. hadronic mass s_3. As can be seen from Fig. 2, in either case (work [1] or work [2]), the NLL QCD contributions to the decay width are large in general.

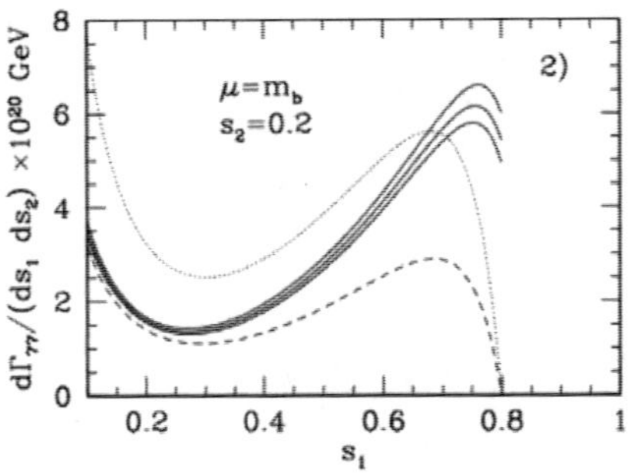

Figure 2. Plot taken from Ref. [2] illustrating the double differential decay width, based on the operator $\mathcal{O}_7$ only, as a function of s_1 for s_2 fixed at $s_2 = 0.2$ and $\mu = m_b$. The dotted, the dashed and the solid lines show the LL result, the NLL when only retaining leading power terms as in Ref. [1] and the full NLL result of the recent paper [2], respectively. Among the three solid lines, the highest, middle and lowest curve correspond to setting $m_s = 400$ MeV, $m_s = 500$ MeV and $m_s = 600$ MeV, respectively.

5. Summary

In these proceedings we reviewed the calculation of the set of $O(\alpha_s)$ corrections to the observable $\bar{B} \to X_s\gamma\gamma$ originating from diagrams involving $\mathcal{O}_7$. To perform this calculation, it was necessary to work out diagrams with three particles (s-quark and two photons) and four particles (s-quark, two photons and a gluon) in the final state. From the technical point of view, the calculation was made possible by the use of the Laporta Algorithm to identify the needed master integrals and by applying the differential equation method to solve the master integrals. When calculating the bremsstrahlung corrections, we took into account only terms proportional to the leading power of the hadronic mass. We found that the infrared and collinear singularities cancel when combining the above mentioned approximated version of bremsstrahlung corrections with the virtual corrections. The numerical impact of the NLL corrections is found to be large: for $d\Gamma_{77}/(ds_1\, ds_2)$ the NLL result is approximately 50% smaller than the LL prediction.

Moreover, we compared our present results with those obtained in a recent paper [2], where a massive strange quark was considered to regulate the collinear photon (gluon) configurations, retaining all powers w.r.t. the hadronic mass.

Acknowledgments

I wish to thank to the organisers of the conference TROIA'14 for their efforts to make it very pleasant.

[1] Asatrian H, Greub C, Kokulu A and Yeghiazaryan A 2012 *Phys.Rev.* **D85** 014020 (*Preprint* 1110.1251)
[2] Asatrian H M and Greub C 2014 *Phys.Rev.* **D89** 094028 (*Preprint* 1403.4502)
[3] Misiak M, Asatrian H, Bieri K, Czakon M, Czarnecki A *et al.* 2007 *Phys.Rev.Lett.* **98** 022002 (*Preprint* hep-ph/0609232)
[4] Hurth T and Nakao M 2010 *Ann.Rev.Nucl.Part.Sci.* **60** 645–677 (*Preprint* 1005.1224)
[5] Buras A J 2011 (*Preprint* 1102.5650)
[6] Simma H and Wyler D 1990 *Nucl.Phys.* **B344** 283–316
[7] Reina L, Ricciardi G and Soni A 1997 *Phys.Lett.* **B396** 231–237 (*Preprint* hep-ph/9612387)
[8] Reina L, Ricciardi G and Soni A 1997 *Phys.Rev.* **D56** 5805–5815 (*Preprint* hep-ph/9706253)
[9] Cao J j, Xiao Z j and Lu G r 2001 *Phys.Rev.* **D64** 014012 (*Preprint* hep-ph/0103154)
[10] Kokulu A 2012 (*Preprint* 1207.2397)

Hadron properties from lattice QCD

Constantia Alexandrou

Department of Physics, University of Cyprus, P.O. Box 20537, 1678 Nicosia, Cyprus
Computation-based Science and Technology Research Center, Cyprus Institute, 20 Kavafi Str., Nicosia 2121, Cyprus
NIC, DESY, Platanenallee 6, D-15738 Zeuthen, Germany

E-mail: alexand@ucy.ac.cy

Abstract. We discuss the status of current dyanmical lattice QCD simulations in connection to the emerging results on the low-lying baryon spectrum, excited states of the nucleon and the investigation of the structure of scalar mesons.

1. Introduction

Simulations of lattice QCD are nowadays being performed with dynamical quarks with masses close to their physical ones. Such simulations with physical pions remove the need for a chiral extrapolation, thereby eliminating a significant source of a systematic uncertainty that has proved difficult to quantify in the past. Various fermion discretization schemes are being employed by various collaborations. MILC has recently presented results on the pseudoscalar decay constants using Highly Improved Staggered Quark (HISQ) ensembles with the strange and charm quarks fixed to their physical values and for a range of masses for the two light quarks ($N_f = 2+1+1$) including physical values of the light sea-quark masses (1). The BMW collaboration has produced results on the pion sector using $N_f = 2+1$ $\mathcal{O}(a)$-improved Clover simulations employing HEX smeared links with light quark masses over a range of masses even below the physical pion mass for four lattice spacings (2). A number of other collaborations are using improved Wilson fermions to simulate with physical or near physical values of the two dynamical light quark masses, in some cases including a dynamical strange quark with mass fixed to its physical value. Clover gauge configurations have been produced by the QCDSF and PACS-CS collaborations and pion mass $m_\pi \sim 150$ MeV for $N_f = 2$ (3) and $N_f = 2+1$ with re-weighing to reach the physical pion value (4). The European Twisted Mass Collaboration (ETMC) has also generated $N_f = 2$ gauge configurations using twisted mass fermions including the clover term (5). Using these 'physical ensembles' one has now the possibility to study hadron properties directly.

2. Hadron spectrum

The first quantity that we would like to reproduce from lattice QCD are the masses of the low-lying hadrons. These are extracted from Euclidean correlation functions

$$G(\vec{q}, t_s) = \sum_{\vec{x}_s} e^{-i\vec{x}_s \cdot \vec{q}} \langle J(\vec{x}_s, t_s) J^\dagger(0) \rangle = \sum_{n=0,\cdots,\infty} A_n e^{-E_n(\vec{q})t_s} \overset{t_s \to \infty}{\longrightarrow} A_0 e^{-E_0(\vec{q})t_s} \quad (1)$$

in the large Euclidean time limit, after projecting to zero momentum, $\vec{q} = \vec{0}$.

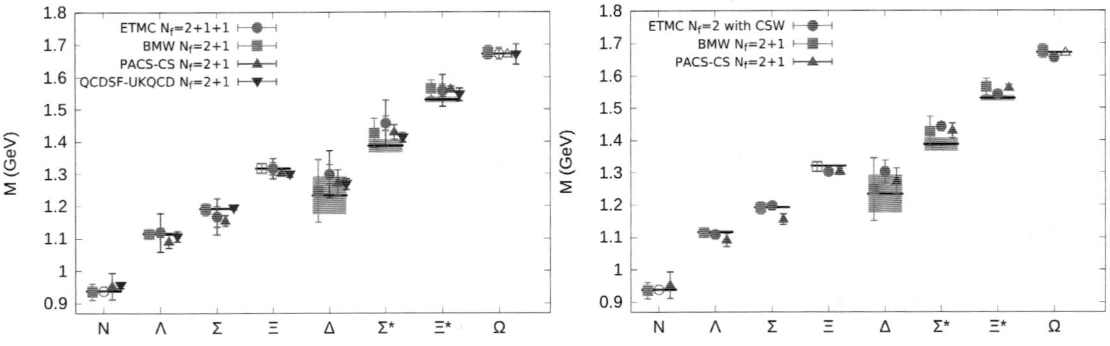

Figure 1. Lattice QCD results on the octet and decuplet baryon masses compared to the experimental values shown by the horizontal bands. Results by ETMC are shown in red circles. Left: using $N_f = 2+1+1$ ensembles after performing a chiral extrapolation (statistical errors are shown in red, whereas the blue error bar includes an estimate of the systematic errors due to the chiral extrapolation (6). Right: for the physical ensemble (7). In both plots we also show results using clover fermions from BMW (8) (magenta squares), from PACS-CS (9) (green triangles), and from QCDSF-UKQCD collaborations (10) using $N_f = 2 + 1$ SLiNC configurations (blue inverted triangles). Open symbols show the baryon mass used as input to the calculations.

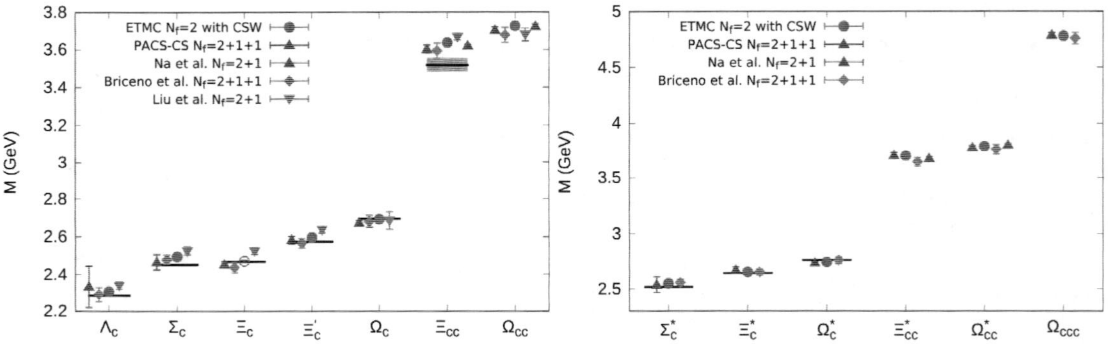

Figure 2. Results by ETMC are shown in red circles for the mass of the spin-1/2 (left) and spin-3/2 (right) charmed baryons for the physical ensemble. Included are results from various hybrid actions with staggered sea quarks from Refs. (11) (purple triangles), (12) (magenta diamonds) and (13) (orange inverted triangles). Results from PACS-CS (14) are shown in green triangles.

In Fig. 2 we show the spectrum of the octet and decuplet baryons. We show two sets of results using twisted mass fermions (TMF). One set is obtained with $N_f = 2 + 1 + 1$ gauge configurations at three lattice spacings, determined using the nucleon mass as $a = 0.094$ fm, 0.082 fm and 0.065 fm. Thus results can be extrapolated to the continuum limit. The continuum results are chirally extrapolated using heavy baryon chiral perturbation theory to leading and next to leading order. We take the difference between the two orders as an estimate of the systematic error due to the chiral extrapolation, which constitutes the biggest systematic error on the results as can be seen by the blue error bars (6). Results obtained at two different lattice volumes and showed no observable effect within our statistics and thus volume corrections were not performed. The other set of TMF results shown in Fig. 2 is obtained using simulations with physical values of the light quark masses (physical ensemble), thus requiring no chiral extrapolation, at one lattice spacing and volume. Both sets agree with experimental values, indicating that finite lattice spacings effects for the physical ensemble are small (7). These results are now much more precise since the chiral extrapolation was not needed. In Fig. 2 we show the corresponding results for the mass of the charmed baryons for the physical ensemble of TMF. As can be seen, the known values of the masses of charmed baryons are reproduced and thus our computation provides a prediction for the yet unmeasured masses. Our preliminary results for the yet unmeasured mass of the Ξ^*_{cc} is 3.700(85) GeV, for the Ω^+_{cc} is 3.726(42) GeV, for Ω^{*+}_{cc} 3.785(49) GeV and for Ω^{++}_{ccc} 4.778(33) GeV.

3. Nucleon excited states

Having successfully computed the mass of the nucleon and the other low-lying baryons from lattice QCD we turn to the question of the excited nucleon spectrum. We limit our discussion here to the two lowest states, namely the $P_{11}(1440)$ positive parity resonance known as the Roper and the negative parity state $S_{11}(1535)$. This is an example where the mass ordering of these states is contrary to the prediction of the constituent quark model where the negative parity state is expected to be lower in mass than the P_{11} state.

The method to study excited states is to use an enlarged basis of interpolating fields and construct a correlation matrix

$$G_{jk}(\vec{q}, t_s) = \sum_{\vec{x}_s} e^{-i\vec{x}_s \cdot \vec{q}} \langle J_j(\vec{x}_s, t_s) J_k^\dagger(0) \rangle \, , j, k = 1, \ldots N \, . \tag{2}$$

Solving the generalized eigenvalue problem (GEVP)

$$G(t) v_n(t; t_0) = \lambda_n(t; t_0) G(t_0) v_n(t; t_0) \rightarrow \lambda_n(t; t_0) = e^{-E_n(t-t_0)} \quad . \tag{3}$$

yields asymptotically the N lowest eigenstates (15).

We have used a variational approach to study the excited states of the nucleon in the positive and negative parity channels (16). Our variational basis consisted of two types of nucleon interpolating fields with different levels of Gaussian smearings.

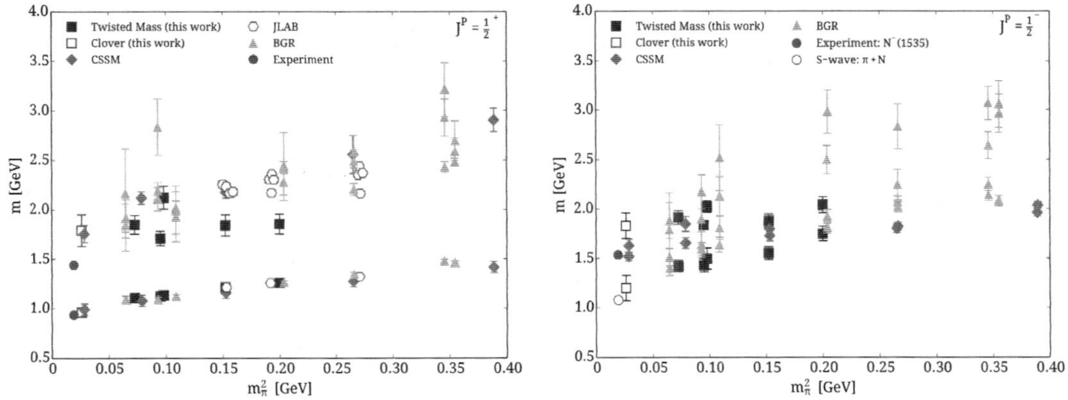

Figure 3. The positive (left) and negative (right) states of the nucleon. Filled and open squares are results using $N_f = 2$ twisted mass and clover fermions respectively. $N_f = 2 + 1$ Clover fermions by the CSSM collaboration (17) (red diamonds), by the Hadron Spectrum Collaboration (18) (open hexagons) using the Chirally Improved Dirac Operator by the Bern-Graz-Regensburg (BGR) collaboration (19) (yellow triangles) are also shown. The CSSM results on the negative parity states are from (20).

We analyzed a total of five ensembles of $N_f = 2$ twisted mass fermions with pion mass in the range of about 270 MeV to 450 MeV and lattice spacing $a = 0.089$ fm determined from the nucleon mass. For this value of the lattice spacing cut-off effects on the mass of the nucleon and hyperons were found to be smaller than the statistical errors. Therefore, we limit ourselves to studying only one lattice spacing. In addition, we analyzed an ensemble of $N_f = 2$ Clover fermions with pion mass $m_\pi \sim 160$ MeV and lattice spacing $a \simeq 0.073$ fm (3). Our results are shown in Fig. 3 for the positive and negative channels. Results from other lattice QCD computations are also included. The first observation is that all lattice results are in reasonable agreement for the ground state energies of both parity channels. The second major observation is that our data for the first excited state of the nucleon in the positive parity channel, although consistent at near physical pion mass with the other lattice calculation at similar pion mass, namely that from the CSSM Collaboration, yield a value that is still higher than the experimentally measured mass for the Roper. Given that our lattice volume is comparable to that of Ref. (17) volume effects can be responsible for the larger values. In the negative parity channel we can clearly see that for all pion masses considered the negative parity ground state is consistent with a πN state in an S-wave. To the statistical accuracy available to us, the first excited negative parity state appears to be converging to $N^-(1535)$, however the errors are too large to draw concrete conclusions. It is also apparent that the results on the higher states are much more spread and carry much larger errors, thus requiring further study.

4. Scalar mesons

Using the variational approach one can study scalar mesons. Our understanding of the light scalar ($J^P = 0^+$) meson sector below 1 GeV is still unclear. The observed mass ordering of the $f_0(980)$ and $a_0(980)$ states appears inverted from what would be naively expected from the conventional quark model. Instead, the interpretation as four-quark states provides an explanation of the inverted mass values of these scalars (21). In this four-quark scenario, the quark content will be given by

$$\begin{aligned}
&\sigma = ud\bar{u}\bar{d}, \quad f_0 = \tfrac{1}{\sqrt{2}}\left(us\bar{u}\bar{s} + ds\bar{d}\bar{s}\right), \\
&a_0^- = ds\bar{u}\bar{s}, \quad a_0^0 = \tfrac{1}{\sqrt{2}}\left(us\bar{u}\bar{s} - ds\bar{d}\bar{s}\right), \quad a_0^+ = us\bar{d}\bar{s} \\
&\kappa^+ = ud\bar{d}\bar{s}, \quad \kappa^0 = ud\bar{u}\bar{s}, \quad \bar{\kappa}^0 = us\bar{u}\bar{d}, \quad \kappa^- = ds\bar{u}\bar{d} \quad .
\end{aligned} \tag{4}$$

Within this interpretation, the mass degeneracy of $f_0(980)$ and $a_0(980)$ is natural and the mass ordering is understandable. The broad width of the σ and κ can be also easier explained since the decay channels to $\pi\pi$ and $K\pi$ respectively are OZI super-allowed.

A number of lattice QCD studies of scalar mesons have been undertaken (22–25). Our ongoing investigation of the low-lying scalar nonet (26) focuses first on the study of the $a_0(980)$ in an effort to shed light on its structure from QCD. Our variational basis consists of the conventional two quark operator, as well as, different types of four quark operators, corresponding to two mesons or a diquark-antidiquark structure. Including the two quark operator inevitably means that there are diagrams which contain closed fermion loops, which makes the study even more challenging. Preliminary results are obtained using a 4 correlation matrix and $N_f = 2+1$ clover gauge configurations generated by the PACS-CS collaboration (27). The lattice size is $32^3 \times 64$ with lattice spacing $a \approx 0.09$ fm. Two ensembles are analyzed corresponding to $m_\pi \approx 300$ MeV, and $m_\pi \approx 150$ MeV. Results are still preliminary and currently we are improving our calculations. Including explicitly scattering states for the $\eta_{ss} + \pi$ and $K + \bar{K}$ type with explicit projection to zero momentum is expected to improve our ability to isolate the $a_0(980)$ state.

5. Conclusions

Simulations at the physical point are now becoming available enabling the computation of the masses of the low-lying hadron without the need of chiral extrapolation. This eliminates an up to now ill-determined systematic error from lattice QCD evaluations. Results on other benchmark quantities such as the nucleon axial charge g_A, and momentum fractions $\langle x\rangle_{u-d}$ are also being computed directly at the physical point (7; 28). We will need high statistics and careful cross-checks to finalize results on these quantities at the physical point. The study of excited states and resonances is under way but many challenges need still to be resolved.

6. Acknowledgments

I would like to thank my close collaborators A. Adbel-Rehim, J. Berlin, M. Dalla Brida, V. Drach, M. Gravina, K. Jansen, Ch. Kallidonis, G. Koutsou, T. Leontiou and M. Wagner for their valuable contributions without which this work would not have been possible. I am also indebted to the QCDSF collaboration (3) for making the clover configurations available to us.

References

[1] A. Bazavov *et al.* [Fermilab Lattice and MILC Collaboration], arXiv:1407.3772 [hep-lat].
[2] S. Durr *et al.*, arXiv:1310.3626 [hep-lat].
[3] G. S. Bali *et al.*, Nucl. Phys. B **866**, 1 (2013).
[4] S. Aoki *et al.* [PACS-CS Collaboration], Phys. Rev. D **81**, 074503 (2010).
[5] A. Abdel-Rehim *et al.*, PoS LATTICE **2013**, 264 (2013).
[6] C. Alexandrou, V. Drach, K. Jansen, C. Kallidonis and G. Koutsou, arXiv:1406.4310.
[7] C. Alexandrou, M. Constantinou, K. Hadjiyiannakou, K. Jansen, C. Kallidonis and G. Koutsou, PoS (Lattice2014) 151.
[8] S. Durr *et al.* Science **322**, 1224 (2008).
[9] S. Aoki *et al.* [PACS-CS Collaboration], Phys. Rev. D **79**, 034503 (2009).
[10] W. Bietenholz *et al.*, Phys. Rev. D **84**, 054509 (2011).
[11] H. Na and S. A. Gottlieb, PoS LAT **2007**, 124 (2007); H. Na and S. Gottlieb, PoS LATTICE **2008**, 119 (2008).
[12] R. A. Briceno, H. W. Lin and D. R. Bolton, Phys. Rev. D **86**, 094504 (2012).
[13] L. Liu, H. W. Lin, K. Orginos and A. Walker-Loud, Phys. Rev. D **81**, 094505 (2010).
[14] Y. Namekawa *et al.* [PACS-CS Collaboration], Phys. Rev. D **87**, 094512 (2013).
[15] M. Luscher and U. Wolff, Nucl. Phys. B **339**, 222 (1990).
[16] C. Alexandrou, T. Korzec, G. Koutsou and T. Leontiou, Phys. Rev. D **89**, 034502 (2014).
[17] M. S. Mahbub *et. al*, AIP Conf. Proc. **1441**, 293 (2012).
[18] R. G. Edwards, J. J. Dudek, D. G. Richards, S. J. Wallace, Phys. Rev. D **84**, 074508 (2011).
[19] G. P. Engel *et al.* [BGR Collaboration], Phys. Rev. D **87**, 074504 (2013).
[20] M. S. Mahbub *et al.*, Phys. Rev. D **87**, 094506 (2013).
[21] J. R. Pelez, AIP Conf. Proc. **1606**, 189 (2014).
[22] K. F. Liu, arXiv:0805.3364 [hep-lat].
[23] S. Prelovsek, arXiv:0804.2549 [hep-lat].
[24] C. Bernard, C. E. DeTar, Z. Fu and S. Prelovsek, Phys. Rev. D **76**, 094504 (2007).
[25] M. Wakayama and C. Nonaka, PoS LATTICE **2012**, 276 (2012).
[26] C. Alexandrou *et al.*, JHEP **1304**, 137 (2013).
[27] S. Aoki *et. al*, Phys. Rev. D79, 034503 (2009) arXiv:0807.1661.
[28] C. Alexandrou, M. Constantinou, V. Drach, K. Jansen, C. Kallidonis and G. Koutsou, PoS LATTICE **2013**, 292 (2013).

Journal of Physics: Conference Series **562** (2014) 012008 doi:10.1088/1742-6596/562/1/012008

Constraint Analysis for the Interaction of the Vector-Meson Octet with the Baryon Octet

Y. Ünal, A. Küçükarslan, S. Scherer

Institut für Kernphysik, Johannes Gutenberg-Universität, D-55099 Mainz, Germany,
Çanakkale Onsekiz Mart University, Department of Physics, 17100 Çanakkale, Turkey

E-mail: uenal@kph.uni-mainz.de, yaseminunal@comu.edu.tr

Abstract. We describe a constraint analysis for the interaction of the vector-meson octet with the baryon octet. Applying Dirac's Hamiltonian method, we verify that the standard interaction in terms of two independent SU(3) structures is consistent at the classical level. We argue how the requirement of self consistency with respect to perturbative renormalizability may lead to relations among the renormalized coupling constants of the system.

1. Introduction

In quantum chromodynamics (QCD), the physics of strongly interacting particles is described in terms of quarks and gluons as the dynamical degrees of freedom (DOF). In the low-energy regime, QCD can be approximated by an effective field theory (EFT) using hadronic dynamical DOF. To that end, one writes down the most general Lagrangian consistent with the assumed symmetries of the fundamental theory [1]. In addition, one needs a power-counting scheme for the expansion of physical observables.

The interaction of the pseudoscalar octet (π, K, η) with the baryon octet is largely constrained by spontaneous symmetry breaking [2]. It is also interesting to investigate the coupling of the vector-meson octet to the baryon octet. The situation is more complex, because a Lorentz-invariant description of spin-1 systems introduces unphysical degrees of freedom. Therefore, one imposes constraints which, for an interacting theory, may lead to relations among the coupling constants of the Lagrangian.

We investigate the lowest-order effective Lagrangian for the interaction of the vector-meson octet with the baryon octet by performing a classical Dirac constraint analysis [3]. For the quantized theory we demand that it is perturbatively renormalizable in the sense of effective field theory [4]. For the pure vector-meson sector, such an investigation results in a massive SU(3) Yang-Mills theory [5, 6]. For the interaction of the vector mesons with the baryons we search for additional relations among the coupling constants.

2. Classical Dirac Constraint Analysis

The most general effective Lagrangian for a system of a massive vector-meson octet interacting with a massive baryon octet can be written as

$$\begin{aligned}
\mathcal{L} &= \mathcal{L}_1 + \mathcal{L}_{1/2} + \mathcal{L}_{\rm int} + \cdots, \\
\mathcal{L}_1 &= -\frac{1}{4}V_{a\mu\nu}V_a^{\mu\nu} + \frac{M_V^2}{2}V_{a\mu}V_a^\mu - gf_{abc}\partial_\mu V_{a\nu}V_b^\mu V_c^\nu - \frac{g^2}{4}f_{abc}f_{ade}V_{b\mu}V_{c\nu}V_d^\mu V_e^\nu, \\
\mathcal{L}_{1/2} &= \bar{\Psi}_a\left(i\gamma^\mu\partial_\mu - m\right)\Psi_a, \\
\mathcal{L}_{\rm int} &= i\mathrm{G_F}f_{abc}\bar{\Psi}_a\gamma^\mu\Psi_b V_{c\mu} + \mathrm{G_D}d_{abc}\bar{\Psi}_a\gamma^\mu\Psi_b V_{c\mu},
\end{aligned} \tag{1}$$

where the SU(3) indices a, b, c range from 1 to 8. The three coupling constants g, $\mathrm{G_F}$, and $\mathrm{G_D}$ are dimensionless. In Eq. (1), the ellipses stand for "nonrenormalizable" higher-order interactions, which we assume to be suppressed by powers of some large scale, as well as for interactions with other hadrons. We take the Lagrangian to be invariant under *global* SU(3) transformations. For the vector-meson self interaction, the constraint analysis of Refs. [6, 7] has already been incorporated, leading to a reduction from five independent couplings to one single coupling g.

In the canonical formalism, the momentum field variables conjugate to the field variables are given by

$$\pi_{a\mu} = \frac{\partial\mathcal{L}}{\partial\dot{V}_a^\mu} = \frac{\partial\mathcal{L}_1}{\partial\dot{V}_a^\mu}, \qquad \Pi_{\Psi a} = \frac{\partial^R\mathcal{L}}{\partial\dot{\Psi}_a} = \frac{\partial^R\mathcal{L}_{1/2}}{\partial\dot{\Psi}_a}, \qquad \Pi_{\Psi^\dagger a} = \frac{\partial^L\mathcal{L}}{\partial\dot{\Psi}_a^\dagger} = \frac{\partial^L\mathcal{L}_{1/2}}{\partial\dot{\Psi}_a^\dagger}. \tag{2}$$

Because ψ and $\psi^\dagger$ are anti-commuting variables, we consider the convention of taking fermionic derivatives from the left for the adjoint field and from the right for the field. Using these relations, we immediately see that the velocities cannot be expressed in terms of the momenta. In this case, we cannot go from the configuration space of the system to its phase space. To define the Hamiltonian of the system, we introduce three so-called primary constraints [3],

$$\theta_{Va}^1 = \pi_{a0} + gf_{abc}V_{b0}V_{c0} \approx 0, \qquad \chi_{\Psi a} = \Pi_{\Psi a} - \frac{i}{2}\Psi_a^\dagger \approx 0, \qquad \chi_{\Psi^\dagger a} = \Pi_{\Psi^\dagger a} + \frac{i}{2}\Psi_a \approx 0. \tag{3}$$

Here, a relation such as $\theta_{Va}^1 \approx 0$ denotes a weak equation in Dirac's sense [3], namely that one must not use one of these constraints before working out a Poisson bracket. In total, Eq. (3) amounts to 8 constraints for the vector mesons and $8\cdot 4 + 8\cdot 4 = 64$ constraints for the baryons. We introduce unknown Lagrange multiplier functions $(\lambda_a, \lambda_a^\dagger, \lambda_{Va})$, define a constraint Hamiltonian (density) $\mathcal{H}_c$, and construct the total Hamiltonian (density) in terms of a Legendre transformation,

$$\mathcal{H}_c = \lambda_a\chi_{\Psi a} + \lambda_a^\dagger\chi_{\Psi^\dagger a} + \lambda_{Va}\theta_{Va}^1, \qquad \mathcal{H}_T = \mathcal{H}_1 + \mathcal{H}_{1/2} + \mathcal{H}_{\rm int} + \mathcal{H}_c. \tag{4}$$

Table 1. Counting the DOF for the free vector, Dirac, and interacting theories, respectively.

Case	Total DOF	Constraints	Physical DOF
Free vector fields	64	16	48
Free Dirac fields	128	64	64
Interacting theory	192	80	112

The requirement that Eqs. (3) have to be zero throughout all time results in

$$\begin{aligned}
\{\theta^1_{Va}, H_T\} &= i\mathrm{G_F} f_{abc}\Psi^\dagger_b\Psi_c - \mathrm{G_D} d_{abc}\Psi^\dagger_b\Psi_c - \partial_i\pi^i_a - M^2_V V_{a0} \\
&\quad + g f_{abc}\pi_{bi}V^i_c - g f_{abc}\partial_i V^i_b V_{c0} - g f_{abc}\partial_i V_{b0}V^i_c \\
&= \theta^2_{Va} \approx 0, \qquad (5)\\
\{\chi_{\Psi a}, H_T\} &= i\partial_i\Psi^\dagger_a\gamma^0\gamma^i + m\Psi^\dagger_a\gamma^0 - ... - i\lambda^\dagger_a = 0, \qquad (6)\\
\{\chi_{\Psi^\dagger_a}, H_T\} &= -i\partial_i\Psi_a\gamma^0\gamma^i + m\Psi_a\gamma^0 + ... - i\lambda_a = 0, \qquad (7)
\end{aligned}$$

where $H_T = \int d^3x\,\mathcal{H}_T$. Equation (5) is a so-called secondary constraint, and, therefore, we obtain 8 additional constraints. Evaluating the Poisson bracket of θ^2_{Va} and H_T results in an equation for the Lagrange multiplier λ_{Va}. From Eqs. (6) and (7) we can solve for the Lagrange multipliers $\lambda^\dagger_a$ and λ_a, respectively. At this stage, we have solved for all the Lagrange multipliers and have generated the correct number of constraints. As a result of Dirac's constraint analysis, at the classical level we have a self-consistent theory with the correct number of constraints and thus the correct number of physical DOF without any relation among the couplings.

3. Perturbative Renormalizability

Pertubative renormalizability in the sense of EFT requires that the ultraviolet divergences of loop diagrams can be absorbed in the redefinition of the masses, coupling constants, and fields of the *most general* effective Lagrangian. As in Refs. [5, 6, 7, 8], we expect additional relations among the coupling constants due to the perturbative renormalizability condition. Redefining the fields and parameters, the counter-term Lagrangian reads

$$\begin{aligned}
\mathcal{L}_{\rm ct} = &-\frac{1}{4}\delta Z_V V_{a\mu\nu}V^{\mu\nu}_a + \frac{1}{2}\delta\{M^2_V\}V_{a\mu}V^\mu_a - \frac{1}{4}\delta\{g^2\}f_{abc}f_{ade}V_{b\mu}V_{c\nu}V^\mu_d V^\nu_e - \delta\{g\}f_{abc}\partial_\mu V_{a\nu}V^\mu_b V^\nu_c \\
&+ i\delta Z_\Psi\bar\Psi_a\gamma^\mu\partial_\mu\Psi_a - \delta\{m\}\bar\Psi_a\Psi_a + i\delta\{\mathrm{G_F}\}f_{abc}\bar\Psi_a\gamma^\mu\Psi_b V_{c\mu} + \delta\{\mathrm{G_D}\}d_{abc}\bar\Psi_a\gamma^\mu\Psi_b V_{c\mu}.
\end{aligned} \qquad (8)$$

By comparing the expression for δg obtained from the VVV- and VVVV-vertex functions, we can test whether there are relations among the coupling constants g, $\mathrm{G_F}$, and $\mathrm{G_D}$. The "worst case" would be if all couplings were independent. A different scenario is a type of universality, relating g to a linear combination of $\mathrm{G_F}$ and $\mathrm{G_D}$. Finally, all couplings could be expressible in terms of a single parameter.

4. Conclusion

At the classical level, the standard lowest-order SU(3)-invariant Lagrangian involving three independent coupling constants g, $\mathrm{G_F}$, and $\mathrm{G_D}$ is a self-consistent starting point. This was explicitly shown using Dirac's method. Whether the requirement of perturbative renormalizability implies additional constraints among the couplings remains to be seen.

References

[1] S. Weinberg, Physica A **96**, 327 (1979).
[2] S. Scherer, Adv. Nucl. Phys. **27**, 277 (2003).
[3] P. A. M. Dirac, *Lectures on Quantum Mechanics*, Dover, Mineola, N.Y., 2001.
[4] S. Weinberg, *The Quantum Theory Of Fields. Vol. 1: Foundations*, Cambridge University Press, Cambridge, England, 1995, Chap. 12.
[5] D. Djukanovic, J. Gegelia, and S. Scherer, Int. J. Mod. Phys. A **25**, 3603 (2010).
[6] A. Neiser, *Effective Field Theories for Vector Particles and Constraint Analysis*, Diploma Thesis, Mainz, 2011.
[7] A. Neiser and S. Scherer, in *Light Meson Dynamics*, Mini Proceedings of the Workshop, Mainz, Germany, 2014, edited by W. Gradl, P. Masjuan, M. Ostrick, and S. Scherer, arXiv:1403.6380 [hep-ph].
[8] D. Djukanovic, M. R. Schindler, J. Gegelia, G. Japaridze, and S. Scherer, Phys. Rev. Lett. **93**, 122002 (2004).

First lattice calculation of charmed hadrons' electromagnetic form factors

K U Can[1,2], G Erkol[1], B Isildak[1], M Oka[2] and T T Takahashi[3]

[1] Department of Natural and Mathematical Sciences, Faculty of Engineering, Ozyegin University, Nisantepe Mah. Orman Sok. No:13, Alemdag 34794 Cekmekoy, Istanbul Turkey
[2] Department of Physics, H-27, Tokyo Institute of Technology, Meguro, Tokyo 152-8551 Japan
[3] Gunma National College of Technology, Maebashi, Gunma 371-8530, Japan

E-mail: utku.can@th.phys.titech.ac.jp

Abstract. Electromagnetic form factors of D and D* mesons and Ξ_{cc}, Σ_c, Ω_c and Ω_{cc} baryons are calculated in 2+1 flavor lattice QCD. As a by product of this calculation electric/magnetic charge radii and magnetic moments are extracted. Compared to the PDG values of the light-sector, i.e. pion and proton, charmed hadron results are systematically smaller.

1. Introduction

Electromagnetic form factors are good instruments to probe the internal structure of the hadrons and extract information about their sizes and shapes. Lattice discretisation of QCD provides an *ab initio* scheme to perform non-perturbative calculations to extract form factors. In this proceedings we report our work on the electromagnetic form factors of charmed hadrons, namely D and D* mesons [1] and Ξ_{cc}, Σ_c, Ω_c and Ω_{cc} baryons [2]. Baryon calculations are expanded and improved upon previous Ξ_{cc} calculation [3].

2. Form Factors

Matrix element of the vector current can be written in the form

$$\langle D(p')|V_\mu(q)|D(p)\rangle = (p+p')_\mu \left[e_c F^c(Q^2) + e_q F^q(Q^2)\right] \tag{1}$$

for the spin-0 D meson. As for the spin-1 D* meson, we have

$$\begin{aligned}\langle D^*(p',s')|V_\mu(q)|D^*(p,s)\rangle = \epsilon'^*_\tau(p',s')\Big\{&\tilde{G}_1(Q^2)(p^\mu+p^{\mu\prime})g^{\tau\sigma}\\ &+\tilde{G}_2(Q^2)(g^{\mu\sigma}q^\tau - g^{\mu\tau}q^\sigma) - \tilde{G}_3(Q^2)q^\tau q^\sigma\frac{(p^\mu+p^{\mu\prime})}{2m_{D^*}^2}\Big\}\epsilon_\sigma(p,s),\end{aligned} \tag{2}$$

where ϵ and ϵ' are the polarization vectors of the initial and final vector mesons, respectively. The form factors $\tilde{G}_{1,2,3}$ can be arranged in terms of Sachs electric form factor as follows [4]:

$$\begin{aligned}&F_C(Q^2) = \tilde{G}_1(Q^2) + \frac{2}{3}\eta F_Q(Q^2),\\ &F_M(Q^2) = \tilde{G}_2(Q^2)\\ &F_Q(Q^2) = \tilde{G}_1(Q^2) - \tilde{G}_2(Q^2) + (1+\eta)\tilde{G}_3(Q^2), \quad \text{where } \eta = Q^2/4m_{D^*}^2.\end{aligned} \tag{3}$$

In the baryon sector, matrix element of the vector current interaction of a spin-1/2 baryon is written as

$$\langle \mathcal{B}(p)|V_\mu|\mathcal{B}(p')\rangle = \bar{u}(p)\left[\gamma_\mu F_{1,\mathcal{B}}(q^2) + i\frac{\sigma_{\mu\nu}q^\nu}{2m_\mathcal{B}}F_{2,\mathcal{B}}(q^2)\right]u(p),$$

where $q_\mu = p'_\mu - p_\mu$ is the transferred four-momentum. Here $u(p)$ denotes the Dirac spinor for the baryon with four-momentum p^μ and mass $m_\mathcal{B}$. The Sachs electric and magnetic form factors are related to the Dirac, $F_{1,\mathcal{B}}(q^2)$, and Pauli, $F_{2,\mathcal{B}}(q^2)$, form factors by the relations

$$G_{E,\mathcal{B}}(q^2) = F_{1,\mathcal{B}}(q^2) + \frac{q^2}{4m_\mathcal{B}^2}F_{2,\mathcal{B}}(q^2), \quad G_{M,\mathcal{B}}(q^2) = F_{1,\mathcal{B}}(q^2) + F_{2,\mathcal{B}}(q^2).$$

3. Lattice Formulation and Setup

One can extract the Sachs form factors in the large Euclidean time limit, *viz.*, $t_2 - t_1$ and $t_1 \gg a$, by calculating the following ratio

$$R(t_2,t_1;\mathbf{p}',\mathbf{p};\mathbf{\Gamma};\mu) = \frac{\langle F^{\mathcal{H}\mathcal{V}_\mu\mathcal{H}'}(t_2,t_1;\mathbf{p}',\mathbf{p};\mathbf{\Gamma})\rangle}{\langle F^{\mathcal{H}\mathcal{H}}(t_2;\mathbf{p}';\Gamma_4)\rangle}\left[\frac{\langle F^{\mathcal{H}\mathcal{H}}(t_2-t_1;\mathbf{p};\Gamma_4)\rangle\langle F^{\mathcal{H}\mathcal{H}}(t_1;\mathbf{p}';\Gamma_4)\rangle\langle F^{\mathcal{H}\mathcal{H}}(t_2;\mathbf{p}';\Gamma_4)\rangle}{\langle F^{\mathcal{H}\mathcal{H}}(t_2-t_1;\mathbf{p}';\Gamma_4)\rangle\langle F^{\mathcal{H}\mathcal{H}}(t_1;\mathbf{p};\Gamma_4)\rangle\langle F^{\mathcal{H}\mathcal{H}}(t_2;\mathbf{p};\Gamma_4)\rangle}\right]^{1/2}, \quad (4)$$

where $\langle F^{\mathcal{H}\mathcal{H}}(t;\mathbf{p};\Gamma_4)\rangle$ and $\langle F^{\mathcal{H}\mathcal{V}_\mu\mathcal{H}'}(t_2,t_1;\mathbf{p}',\mathbf{p};\mathbf{\Gamma})\rangle$ are the two-point and three-point correlation functions with t_1 and t_2 being the current insertion and the sink particle's time slice respectively. In case of baryons $\Gamma_4 \equiv (1+\gamma_4)/2$ and $\Gamma_i = \gamma_i\gamma_5\Gamma_4$, whereas for mesons $\mathbf{\Gamma} = \Gamma_4 \equiv 1$.

Our method of computing the baryon matrix element in Eq. 4, which was employed to extract the nucleon electromagnetic form factor, follows closely that of Ref. [5]. We extract the form factors $G_{E,\mathcal{B}}(q^2)$ and $G_{M,\mathcal{B}}(q^2)$ from Eq. (4) by choosing appropriate combinations of Lorentz direction μ and projection matrices Γ:

$$R(t_2,t_1;\mathbf{0},-\mathbf{q};\Gamma_4;\mu=4) \xrightarrow[t_2-t_1\gg a]{t_1\gg a} \left[\frac{(E_\mathcal{B}+m_\mathcal{B})}{2E_\mathcal{B}}\right]^{1/2}\mathbf{G}_{\mathbf{E},\mathcal{B}}(\mathbf{q}^2), \quad (5)$$

$$R(t_2,t_1;\mathbf{0},-\mathbf{q};\Gamma_j;\mu=i) \xrightarrow[t_2-t_1\gg a]{t_1\gg a} \left[\frac{1}{2E_\mathcal{B}(E_\mathcal{B}+m_\mathcal{B})}\right]^{1/2}\epsilon_{ijk}\,q_k\,\mathbf{G}_{\mathbf{M},\mathcal{B}}(\mathbf{q}^2). \quad (6)$$

Interested reader can consult to the Ref. [2] and references therein for the details.

For D meson, the ratio in Eq. (4) reduces to

$$R(t_2,t_1;\mathbf{0},\mathbf{p};0) \xrightarrow[t_2-t_1\gg a]{t_1\gg a} \frac{(E_D+m_D)}{2\sqrt{E_D\,m_D}}[e_cF^c(Q^2)+e_qF^q(Q^2)], \quad (7)$$

where m_D and E_D are the mass and the energy of the initial baryon. As for the D^* meson the ratio in Eq. (4) has different components for the different polarization vector directions of the initial and final particles and reduces to

$$R^0_{ii} = \frac{p_i^2}{3m_{D^*}\sqrt{E_{D^*}m_{D^*}}}F_Q(Q^2) + \frac{E_{D^*}+m_{D^*}}{2\sqrt{E_{D^*}m_{D^*}}}F_C(Q^2) \quad (8)$$

$$R^0_{jj}\Big|_{j\neq i} = -\frac{p_i^2}{6m_{D^*}\sqrt{E_{D^*}m_{D^*}}}F_Q(Q^2) + \frac{E_{D^*}+m_{D^*}}{2\sqrt{E_{D^*}m_{D^*}}}F_C(Q^2), \quad (9)$$

where i and j are the polarization directions. In order to single out the electric form factor we compute

$$\frac{1}{3}\sum_{i=1,2,3}R^0_{ii}(t_2,t_1;\mathbf{0},p_j;0) \xrightarrow[t_2-t_1\gg a]{t_1\gg a} \frac{(E_{D^*}+m_{D^*})}{2\sqrt{E_{D^*}\,m_{D^*}}}[e_cF^{*c}(Q^2)+e_qF^{*q}(Q^2)]. \quad (10)$$

Table 1: Extrapolated results of charge radii and magnetic moments.

	$\langle r^2_{E,D}\rangle$	$\langle r^2_{E,D^*}\rangle$	$\langle r^2_{E,\Sigma_c^{++}}\rangle$	$\langle r^2_{E,\Xi_{cc}^{++}}\rangle$	$\langle r^2_{E,\Omega_{cc}^{+}}\rangle$	$\langle r^2_{M,\Sigma_c^{++}}\rangle$	$\langle r^2_{M,\Sigma_c^{0}}\rangle$	$\langle r^2_{M,\Omega_c^{0}}\rangle$
Lin. Fit	0.138(13)	0.185(24)	0.192(22)	0.136(8)	0.032(6)	0.410(81)	0.377(75)	0.297(33)
Quad. Fit	0.152(26)	0.192(43)	0.234(37)	0.165(12)	0.043(11)	0.696(153)	0.650(126)	0.354(54)
	$\langle r^2_{M,\Xi_{cc}^{+}}\rangle$	$\langle r^2_{M,\Omega_{cc}^{+}}\rangle$	$\mu_{\Sigma_c^{++}}$	$\mu_{\Sigma_c^{0}}$	$\mu_{\Omega_c^{0}}$	$\mu_{\Xi_{cc}^{+}}$	$\mu_{\Omega_{cc}^{+}}$	
Lin. Fit	0.135(10)	0.135(11)	1.569(253)	-0.852(133)	-0.608(45)	0.411(15)	0.405(13)	
Quad. Fit	0.154(19)	0.148(21)	2.220(505)	-1.073(269)	-0.639(88)	0.425(29)	0.413(24)	

Details can be found in Ref. [1] and references therein.

We run our simulations on $32^3 \times 64$, 2+1-flavor configurations generated by the PACS-CS Collaboration [6] using the Clover quark action and the Iwasaki gauge action. We use the gauge configurations at $\beta = 1.90$ with the Clover coefficient $c_{SW} = 1.715$, which give a lattice spacing of $a = 0.0907(13)$ fm ($a1 = 2.176(31)$ GeV). The simulations are carried out with four different hopping parameters for the sea and the u, d valence quarks, κ_{sea}, $\kappa_{u,d}$=0.13700, 0.13727, 0.13754 and 0.13770, which correspond to pion masses of approximately 702, 570, 411, and 296 MeV. The hopping parameter for the strange valance is fixed to κ_s=0.1364, same as the sea-quark's. We use Clover action for charm quarks also, with $c_{SW} = 1/u_0^3$ where u_0 is the average link, and with hopping parameters $\kappa_c^{\mathrm{meson}} = 0.1224$ and $\kappa_c^{\mathrm{baryon}} = 0.1246$. Multiple gaussian-smeared source – wall-smeared sink pairs with $t = 12a$ separation have been used to increase the statistics.

4. Results and Discussion

Charge radii can be extracted from the slope of the form factors at zero momentum transfer by $\langle r^2\rangle = -\frac{6}{G(0)}\frac{d}{dQ^2}G(Q^2)|_{Q^2=0}$. We choose to model the form factors by $G(Q^2) = G(0)/(1+Q^2/\Lambda^2)^a$, where $G(Q^2)$ denotes the form factor and Λ is a fit parameter. We take $a = 1$ for mesons and $a = 2$ for the baryons. The charge radii, then, is calculated via, $\langle r^2\rangle = 6a/\Lambda^2$. Magnetic moments of baryons are obtained from the magnetic form factor at zero momentum transfer by evaluating, $\mu_{\mathcal{B}} = G_M(0)\,(e/2m_{\mathcal{B}}) = G_M(0)\,(m_N/m_{\mathcal{B}})\,\mu_N$.

Plateau, form factor and chiral extrapolation plots can be found in Refs. [2, 1], here we only quote the values extrapolated to the physical point. We consider two fit forms, one being linear in m_π^2, $f(m_\pi^2)_{lin} = a + bm_\pi^2$ and the other being quadratic as, $f(m_\pi^2)_{quad} = a + bm_\pi^2 + c(m_\pi^2)^2$, where a, b and c are fit parameters. The results are compiled in Table 1.

In contrast to their light counterparts, i.e. $< r^2_{\pi,E} > = 0.452$ fm^2 and $< r^2_{p,E} > =0.770$ fm^2 [7], charmed hadron results are observed to be smaller. Comparison of individual quark sector contributions given in Refs. [2, 1] shows that the decrease in values is mainly due to the small contribution of the charm quark. In the baryon sector, comparison of Ξ_{cc} and Ω_{cc} shows that replacing a d-quark by an s-quark seems to have a negligible effect on the magnetic charge radius. Σ_c^{++} has the largest electric/magnetic charge radius amongst others. Peculiar m_π^2 dependence of Ω_c and Ω_{cc} is a possible indication of the sea-quark effects [1].

Magnetic moments are dominantly determined by the doubly represented quarks and individual quark sector study shows that light and heavy quark spins are anti-aligned [1]. Compared to the experimental result, $\mu_p = 2.793\mu_N$ [7], the values are systematically smaller with Σ_c^{++} being closest to the proton's value. In contrast to the charge radii, magnetic moments of Ω_c and Ω_{cc} seems independent of the sea-quark effects. Lattice results reproduce the signs of magnetic moments correctly but underestimates the values compared to the other models [1]. For detailed discussions of our results we refer the reader to Refs. [2, 1].

5. Conclusions

We have calculated the electromagnetic form factors of the D, D* mesons and Σ_c, Ξ_{cc}, Ω_c and Ω_{cc} baryons on $2+1$-flavor lattices and as a by product extracted the electric and magnetic charge radii and the magnetic moments. Comparison with pion/nucleon data shows that charmed hadrons are more compact. This is believed to be due to the valance charm quarks. Magnetic moments are dominantly determined by the doubly represented quarks. Signs match with other models even though values are underestimated. Simulations on *physical-point* configurations are ongoing.

Acknowledgements

All the numerical calculations in this work were performed on National Center for High Performance Computing of Turkey (Istanbul Technical University) under project number 10462009. The unquenched gauge configurations employed in our analysis were generated by PACS-CS collaboration [6]. We used a modified version of Chroma software system [8]. This work is supported in part by The Scientific and Technological Research Council of Turkey (TUBITAK) under project numbers 110T245, 114F261 and in part by KAKENHI under Contract Nos. 22105503, 24540294 and 22105508.

References

[1] Can K, Erkol G, Oka M, Ozpineci A and Takahashi T 2013 *Physics Letters* B **719** 103 – 109
[2] Can K, Erkol G, Isildak B, Oka M and Takahashi T 2014 *J. High Energy Physics* JHEP05(2014)125
[3] Can K, Erkol G, Isildak B, Oka M and Takahashi T 2013 *Physics Letters* B **726** 703 – 709
[4] Brodsky S J and Hiller J R 1992 *Phys.Rev.* D **46** 2141 – 2149
[5] Alexandrou C, Brinet M, Carbonell J, Constantinou M, Harraud P *et al.* 2011 *Phys.Rev.* D **83** 094502
[6] Aoki S, Ishikawa K I, Ishizuka N, Izubuchi T, Kadoh D, Kanaya K, Kuramashi Y, Namekawa Y, Okawa M, Taniguchi Y, Ukawa A, Ukita N and Yoshié T (PACS-CS Collaboration) 2009 *Phys. Rev.* D **79** 034503
[7] Olive K A *et al.* (Particle Data Group) 2014 *Chin. Phys.* C **38** 090001
[8] Edwards R G and Joo B (SciDAC Collaboration, LHPC Collaboration, UKQCD Collaboration) 2005 *Nucl.Phys.Proc.Suppl.* **140** 832

4th International Hadron Physics Conference (TROIA'14) IOP Publishing
Journal of Physics: Conference Series **562** (2014) 012010 doi:10.1088/1742-6596/562/1/012010

Heavy Mesons in Nuclear Matter and Nuclei

Laura Tolos[1,2], Daniel Cabrera[2], Carmen Garcia-Recio[3], Raquel Molina[4], Juan Nieves[5], Eulogio Oset[5], Angels Ramos[6], Olena Romanets[7], Lorenzo Luis Salcedo[3] and Juan M. Torres-Rincon[8]

[1]Instituto de Ciencias del Espacio (IEEC/CSIC), Campus Universitat Autònoma de Barcelona, Facultat de Ciències, Torre C5, E-08193 Bellaterra (Barcelona), Spain

[2]Institut für Theoretische Physik and Frankfurt Institute for Advanced Studies, Johann Wolfgang Goethe University, Ruth-Moufang-Str. 1, 60438 Frankfurt am Main, Germany

[3]Departamento de Física Atómica, Molecular y Nuclear, and Instituto Carlos I de Física Teórica y Computacional, Universidad de Granada, E-18071 Granada, Spain

[4]Department of Physics, Columbian College of Arts & Sciences, Corcoran Hall, 725 21st Street NW, Washington, DC 20052, USA

[5]Instituto de Física Corpuscular (centro mixto CSIC-UV), Institutos de Investigación de Paterna, Aptdo. 22085, 46071, Valencia, Spain

[6]Departament d'Estructura i Constituents de la Matèria, Universitat de Barcelona, Martí i Franqués 1, 08028 Barcelona, Spain

[7]KVI, University of Groningen, Zernikelaan 25, 9747AA Groningen, The Netherlands

[8]Subatech, UMR 6457, IN2P3/CNRS, Université de Nantes, École de Mines de Nantes, 4 rue Alfred Kastler 44307, Nantes, France

E-mail: `tolos@ice.csic.es`

Abstract. Heavy mesons in nuclear matter and nuclei are analyzed within different frameworks, paying a special attention to unitarized coupled-channel approaches. Possible experimental signatures of the properties of these mesons in matter are addressed, in particular in connection with the future FAIR facility at GSI.

1. Introduction

Over the last decades, matter under extreme conditions of density and temperature has been the subject of study in order to address fundamental aspects of the strong interaction. This study is intimately linked to several experimental programs, such as SIS/GSI, RHIC/BNL, LHC/CERN project and the forthcoming PANDA and CBM experiments at FAIR. In this context, the properties of hadrons with strange and charm content in hot and dense matter are of particular interest. Whereas the properties of mesons with strangeness in dense matter have been analyzed in connection to the study of exotic atoms as well as the analysis of heavy-ion collisions, the properties of mesons with charm are under scrutiny and will play an important role at FAIR.

In this talk we review different approaches to obtain the in-medium properties of the strange and open-charm mesons in nuclear matter and nuclei, paying a special attention to coupled-channels unitarized methods. Several experimental scenarios are analyzed, such as heavy-ion collision data on strange pseudoscalar mesons, the photoproduction of strange vector mesons, the formation of D-mesic nuclei and the D-meson propagation from RHIC to FAIR energies.

2. Strange mesons in matter

2.1. Strange pseudoscalar mesons: $\bar{K}$ in matter

Early works based on relativistic mean-field calculations [1] obtained very deep potentials of a few hundreds of MeVs at saturation density ρ_0 for $\bar{K}$ in matter, in line with the analysis of data on antikaonic atoms using phenomenological models [2]. However, antikaonic-atom data tests matter at the surface of the nucleus and, therefore, do not really provide a suitable constraint on the antikaon-nucleus potential at saturation density.

Later approaches on unitarized theories in coupled channels based on the chiral approach [3, 4] or on meson-exchange potentials [5, 6] obtain a much less attractive potential. In these schemes, the attraction is a consequence of the modified s-wave $\Lambda(1405)$ resonance in the medium due to Pauli blocking [7] together with the self-consistent consideration of the $\bar{K}$ self-energy [3] and the inclusion of self-energies of the mesons and baryons in the intermediate states [4]. Attraction of the order of -50 MeV at normal nuclear matter density is then reached [4, 5, 6]. Moreover, higher-partial waves beyond s-wave [8, 9, 10, 11] become essential for relativistic heavy-ion experiments at beam energies below 2 GeV per nucleon.

One of the latest unitarized approaches for calculating the $\bar{K}$ self-energy in symmetric nuclear matter at finite temperature is that of Refs. [10, 11]. In this model, the $\bar{K}$ self-energy and, hence, the spectral function are obtained from the in-medium $\bar{K}$-nucleon interaction in $s-$ and $p-$waves within a chiral unitary approach. The evolution of the $\bar{K}$ spectral function with density and temperature is shown in the l.h.s. of Fig. 1, defined as

$$S_{\bar{K}}(q_0,\vec{q},T) = -\frac{1}{\pi}\frac{\mathrm{Im}\,\Pi_{\bar{K}}(q_0,\vec{q},T)}{\mid q_0^2-\vec{q}^{\,2}-m_{\bar{K}}^2-\Pi_{\bar{K}}(q_0,\vec{q},T)\mid^2}\,, \tag{1}$$

where $\Pi_{\bar{K}}(q_0,\vec{q},T)$ is the $\bar{K}$ self-energy. The $\bar{K}$ spectral function shows a broad peak that results from a strong mixing between the quasi-particle peak and the $\Lambda(1405)N^{-1}$ and $Y(=\Lambda,\Sigma,\Sigma^*)N^{-1}$ p-wave excitations. Temperature and density soften the p-wave contributions to the spectral function at the quasi-particle energy.

From heavy-ion collisions there has been a lot of activity aiming at extracting the properties of $\bar{K}$ in a dense and hot environment [12]. Initial studies have addressed the $\bar{K}$ production in nucleus-nucleus collisions at SIS energies using a transport model with $\bar{K}$ that were dressed with the Juelich meson-exchange model [13] or analyzing multiplicity ratios [14]. However, the question that still remains is to what extend the properties of $\bar{K}$ mesons are modified in matter.

2.2. Vector mesons with strangeness: $\bar{K}^*$ in matter

With regard to strange vector mesons in the nuclear medium, very little discussion has been made about their properties. Recently, the $\bar{K}^*$ self-energy in symmetric nuclear matter has been obtained within the hidden gauge formalism of Ref. [15]. Two sources for the modification of the $\bar{K}^*$ s-wave self-energy emerged in nuclear matter: one associated to the decay mode $\bar{K}\pi$ modified by nuclear medium effects on the π and $\bar{K}$ mesons (which accounts for the $\bar{K}^*N \to \bar{K}N, \pi Y, \bar{K}\pi N, \pi\pi Y\ldots$ processes, with $Y = \Lambda,\Sigma$); and a second one linked to the interaction of the $\bar{K}^*$ with the nucleons in the medium (which accounts for the direct quasi-elastic process $\bar{K}^*N \to \bar{K}^*N$, as well as other absorption channels involving vector mesons and baryons, $\bar{K}^*N \to VB$), coming from a unitarized coupled-channel calculation. Two resonances are generated dynamically, $\Lambda(1783)$ and $\Sigma(1830)$, which can be identified with the experimentally observed states $J^P = 1/2^-$ $\Lambda(1800)$ and the $J^P = 1/2^-$ PDG state $\Sigma(1750)$, respectively [16].

The $\bar{K}^*$ spectral function at zero temperature is given in a similar way as in Eq. (1) and is displayed in the middle panel of Fig. 1 as a function of the meson energy q_0, for zero momentum and different densities up to 1.5 ρ_0. The dashed line refers to the calculation in free space, where only the $\bar{K}\pi$ decay channel contributes, while the other three lines correspond to the

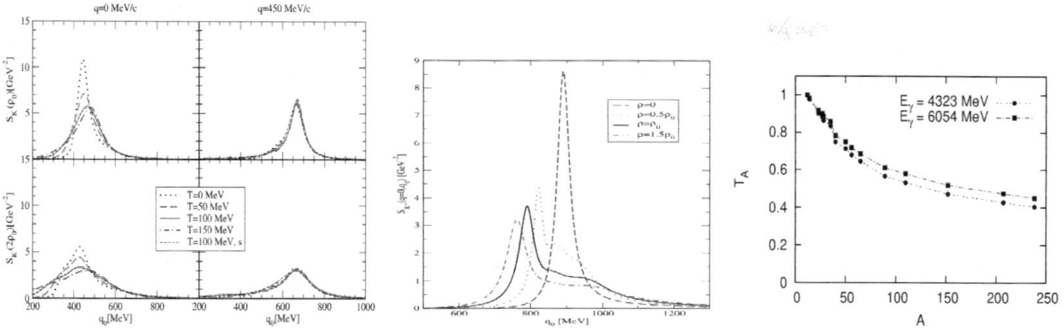

Figure 1. Left: $\bar{K}$ spectral function for different densities, temperatures and momenta [10]. Middle: $\bar{K}^*$ spectral function at zero momentum for different densities [15]. Right: $\bar{K}^*$ transparency ratio [15].

self-consistent calculations, which incorporate the process $\bar{K}^* \to \bar{K}\pi$ in the medium, as well as the quasielastic $\bar{K}^*N \to \bar{K}^*N$ and other $\bar{K}^*N \to VB$ processes. The structures above the quasiparticle peak correspond to the dynamically generated $\Lambda(1783)N^{-1}$ and $\Sigma(1830)N^{-1}$ excitations. Density effects result in a dilution and merging of those resonant-hole states, together with a broadening of the spectral function due to the increase of collisional and absorption processes. What is clear from the present approach, is that the spectral function spread of the $\bar{K}^*$ increases substantially in the medium, becoming at normal nuclear matter density five times bigger than in free space.

In order to test the $\bar{K}^*$ self-energy experimentally, we analyze the normalized nuclear transparency ratio, defined as

$$T_A = \frac{\tilde{T}_A}{\tilde{T}_{^{12}C}}, \quad \text{with} \quad \tilde{T}_A = \frac{\sigma_{\gamma A \to K^+\ K^{*-}\ A'}}{A\,\sigma_{\gamma N \to K^+\ K^{*-}\ N}} \,. \tag{2}$$

It describes the loss of flux of K^{*-} mesons in the nucleus and is related to the absorptive part of the K^{*-}-nucleus optical potential and, thus, to the K^{*-} width in matter. We evaluate the ratio between the nuclear cross sections in heavy nuclei and a light one (^{12}C), T_A, so that other nuclear effects not related to the absorption of the K^{*-} cancel. In the right panel of Fig. 1 we observe a very strong attenuation of the $\bar{K}^*$ survival probability due to the decay $\bar{K}^* \to \bar{K}\pi$ or absorption channels $\bar{K}^*N \to \bar{K}N, \pi Y, \bar{K}\pi N, \pi\pi Y, \bar{K}^*N, \rho Y, \omega Y, \phi Y, \ldots$ with increasing nuclear-mass number A. This is due to the larger path that the $\bar{K}^*$ has to follow before it leaves the nucleus, having then more chances to decay or get absorbed.

3. Open-charm mesons in matter

The medium modifications of mesons with charm, such as D and $\bar{D}$ mesons, have been object of recent theoretical interest due to the consequences for charmonium suppression. A phenomenological estimate based on the quark-meson coupling (QMC) model predicts an attractive D^+-nucleus potential at ρ_0 of $\sim$ -140 MeV [17]. The D-meson mass shift has also been studied using the QCD sum-rule (QSR) approach [18, 19], where a mass shift of -50 MeV at ρ_0 for the D-meson has been suggested [18]. Recent results on QSR rules for open charm mesons can be found in [19]. The mass modification of the D-meson is also addressed using a chiral effective model in hot and dense matter [20], where strong mass shifts were obtained.

With regard to approaches based on coupled-channel dynamics, unitarized methods have been

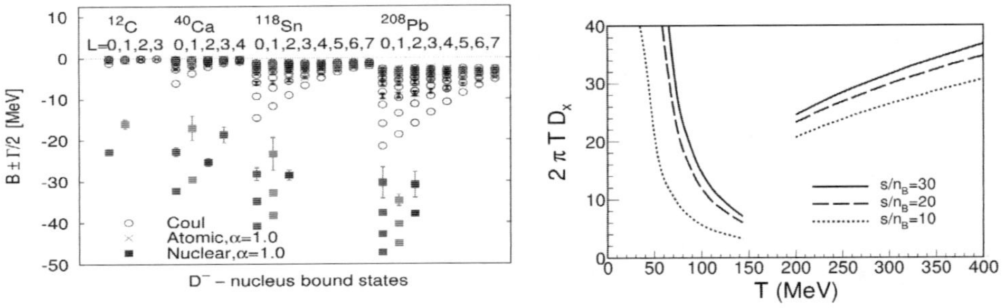

Figure 2. Left: D^--nucleus bound states, with B and Γ the binding energy and width [37]. Right: Spatial diffusion coefficient for D mesons multiplied by $2\pi T$ [39]. For recent updates in the high-temperature phase, see Refs. [40, 41].

applied in the meson-baryon sector with charm content [21, 22, 23, 24, 25, 26, 27, 28, 29, 30, 31], partially motivated by the parallelism between the $\Lambda(1405)$ and the $\Lambda_c(2595)$.

More recently, the implementation of heavy-quark spin symmetry (HQSS) has been considered, which is a proper QCD symmetry that appears when the quark masses, such as the charm mass, become larger than the typical confinement scale. The model generates dynamically resonances with negative parity in all the isospin, spin, strange and charm sectors that one can form from an s-wave interaction between pseudoscalar and vector meson multiplets with $1/2^+$ and $3/2^+$ baryons [32, 33, 34, 35]. Within this model, the self-energies and, hence, spectral functions for D and D^* mesons are obtained self-consistently in a simultaneous manner, as it follows from HQSS, by taking, as bare interaction, an appropiately extended WT interaction [35]. We incorporate Pauli blocking effects and open charm meson self-energies in the intermediate propagators for the in-medium solution [36].

The detection of the in-medium properties of open charm mesons in matter can be addressed, for example, in the possible formation of D mesic nuclei. The QMC model predicted D and $\bar{D}$-meson bound states in ^{208}Pb relying upon an attractive D and $\bar{D}$ -meson potential in the nuclear medium [17]. Within the model that respects HQSS, we obtain that for $\bar{D}$-mesic nuclei [37], not only D^- (as seen in Fig. 2)) but also $\bar{D}^0$ is bound in nuclei . The spectrum contains states of atomic and of nuclear types for all nuclei for D^- while only nuclear states are present for $\bar{D}^0$ in nuclei. Compared to the pure Coulomb levels, the atomic states are less bound. The nuclear ones are more bound and may present a sizable width. Moreover, nuclear states only exist for low angular momenta. In what respects to D mesons [38], D^0-nucleus states are weakly bound in contrast to previous results using the QMC model. Moreover, those states have significant widths, in particular, for ^{208}Pb. Only D^0-nucleus bound states are possible since the Coulomb interaction prevents the formation of observable bound states for D^+ mesons. The experimental detection of D bound states is, however, a difficult task [38].

Another possibility is the study of transport coefficients for a D meson in a hot dense medium composed of light mesons and baryons, such as it is formed in heavy-ion collisions. One interesting observable is the spatial diffusion coefficient D_x around the phase transition from RHIC to FAIR energies [39] (see Fig. 2 right), with a plausible minimum at the phase transition following an isentropic trajectory from the hadronic to the quark-gluon phase [40, 41].

Acknowledgements
This research was supported by DGI and FEDER funds (FIS2011-28853-C02-02, FIS2011-24149, FIS2011-24154, FPA2010-16963, FPA2013-43425-P) and the Spanish Consolider-Ingenio

2010 Programme CPAN (CSD2007-00042), by Junta de Andalucía Grant No. FQM-225, by Generalitat Valenciana under Contract No. PROMETEO/2009/0090, by the Generalitat de Catalunya under contract 2009SGR-1289, by the EU HadronPhysics3 project Grant Agreement No. 283286, Programme "Together" of the Region Pays de la Loire, the FP7- PEOPLE-2011-CIG under Contract No. PCIG09-GA-2011-291679, BMBF (Germany) under project No. 05P12RFFCQ, and Ramon y Cajal Programme.

References

[1] Schaffner J, Bondorf J and Mishustin I N 1997 *Nucl. Phys.* A **625** 325
[2] Friedman E and Gal A 2007 *Phys. Rep.* **452** 89
[3] Lutz M 1998 *Phys. Lett. B* **426** 12
[4] Ramos A and Oset E 2000 *Nucl. Phys.* A **671** 481
[5] Tolos L, Ramos A, Polls A and Kuo T T S 2001 *Nucl. Phys.* A **690** 547
[6] Tolos L, Ramos A and Polls A 2002 *Phys. Rev.* C **65** 054907
[7] Koch V 1994 *Phys. Lett.* B **337** 7
[8] Tolos L, Ramos A and Oset E 2006 *Phys. Rev.* C **74** 015203
[9] Lutz M F M, Korpa C L and Moller M 2008 *Nucl. Phys.* A **808** 124
[10] Tolos L, Cabrera D and Ramos A 2008 *Phys. Rev.* C **78** 045205
[11] Cabrera D, Tolos L, Aichelin J and Bratkovskaya E *Preprint* 1406.2570 [hep-ph]
[12] Hartnack C, Oeschler H, Leifels Y, Bratkovskaya E L and Aichelin J 2012 *Phys. Rep.* **510** 119
[13] Cassing W, Tolos L, Bratkovskaya E L and Ramos A 2003 *Nucl. Phys.* A **727** 59
[14] Tolos L, Polls A, Ramos A and Schaffner-Bielich J 2003 *Phys. Rev.* C **68** 024903
[15] Tolos L, Molina R, Oset E and Ramos A 2010 *Phys. Rev. C* **82** 045210
[16] Oset E, Ramos A, Garzon E J, Molina R, Tolos L, Xiao C W, Wu J J and Zou B S 2012 *Int. J. Mod. Phys.* E **21** 1230011
[17] Tsushima K, Lu D H, Thomas A W, Saito K and Landau R H 1999 *Phys. Rev.* C **59** 2824; Sibirtsev A, Tsushima K and Thomas AW 1999 *Eur. Phys. J.* A **6** 351
[18] Hayashigaki A 2000 *Phys. Lett.* B **487** 96
[19] Hilger T, Thomas R and Kampfer B 2009 *Phys. Rev.* C **79** 025202; Hilger T, Kampfer B and Leupold S 2011 *Phys. Rev.* C **84** 045202
[20] Mishra A, Bratkovskaya E L, Schaffner-Bielich J, Schramm S and Stöcker H 2004 *Phys. Rev.* C **69** 015202; Kumar A and Mishra A 2010 *Phys. Rev.* C **81** 065204; Kumar A and Mishra A 2011 *Eur. Phys. J.* A **47** 164
[21] Tolos L, Schaffner-Bielich J and Mishra A 2004 *Phys. Rev.* C **70** 025203
[22] Tolos L, Schaffner-Bielich J and Stoecker H 2006 *Phys. Lett.* B **635** 85
[23] Lutz M F M and Kolomeitsev E E 2004 *Nucl. Phys.* A **730** 110
[24] Hofmann J and Lutz M F M 2005 *Nucl. Phys.* A **763** 90; Hofmann J and Lutz M F M 2006 *Nucl. Phys.* A **776** 17
[25] Mizutani T and Ramos A 2006 *Phys. Rev.* C **74** 065201
[26] Tolos L, Ramos A and Mizutani T 2008 *Phys. Rev.* C **77** 015207
[27] Molina R, Gamermann D, Oset E and Tolos L 2009 *Eur. Phys. J.* A **42** 31
[28] Jimenez-Tejero C E, Ramos A, Tolos L and Vidana I 2011 *Phys. Rev.* C **84** 015208
[29] Haidenbauer J , Krein G, Meissner U G and Sibirtsev A 2007 *Eur. Phys. J.* A **33** 107
[30] Haidenbauer J, Krein G, Meissner U G and Tolos L 2011 *Eur. Phys. J* A **47** 18
[31] Wu J J , Molina R, Oset E and Zou B S 2010 *Phys. Rev. Lett.* **105** 232001
[32] Garcia-Recio C, Magas V K, Mizutani T, Nieves J, Ramos A, Salcedo L L and Tolos L 2009 *Phys. Rev.* D **79** 054004
[33] Gamermann D, Garcia-Recio C, Nieves J, Salcedo L L and Tolos L 2010 *Phys. Rev.* D **81** 094016
[34] Romanets O, Tolos L, Garcia-Recio C, Nieves J, Salcedo L L and Timmermans R G E 2012 *Phys. Rev.* D **85** 114032
[35] Garcia-Recio C, Nieves J, Romanets O, Salcedo L L and Tolos L 2013 *Phys. Rev.* D **87** 074034
[36] Tolos L, Garcia-Recio C, and Nieves J 2009 *Phys. Rev.* C **80** 065202
[37] Garcia-Recio C, Nieves J, Salcedo L L and Tolos L 2012 *Phys. Rev.* C **85** 025203
[38] Garcia-Recio C, Nieves J and Tolos L 2010 *Phys. Lett.* B **690** 369
[39] Tolos L and Torres-Rincon J M 2013 *Phys. Rev.* D **88** 074019
[40] Berrehrah H, Gossiaux P B, Aichelin J, Cassing W, Torres-Rincon J M and Bratkovskaya E *Preprint* 1406.5322 [hep-ph]
[41] Ozvenchuk V, Torres-Rincon J M, Gossiaux P B, Tolos L and Aichelin J *Preprint* 1408.4938 [hep-ph]

Journal of Physics: Conference Series **562** (2014) 012011 doi:10.1088/1742-6596/562/1/012011

Application of the maximum entropy method to QCD sum rules

Philipp Gubler

RIKEN, Nishina Center, Hirosawa 2-1, Wako, Saitama, 351-0198, Japan

E-mail: pgubler@riken.jp

Abstract. QCD sum rules have long been used to describe the physical properties of hadrons directly from QCD. While this approach was often quite successful, it also has its limitations, the most important one being the need to introduce some specific ansatz for parametrizing the spectral function. For allowing a more general analysis of the sum rules, a new analysis method based on the maximum entropy method has been introduced [1], and has in the meantime been applied to several channels in various environments. In these proceedings, we will discuss some recent results, which have been obtained with the help of this novel approach.

1. Introduction

QCD sum rules already have a long history of studying various hadronic properties [2, 3]. Based on the operator product expansion (OPE) applied to QCD and relying on the analytic properties of hadronic correlators, they provide a class of weighted integrals (the sum rules) over the spectral function, which contain the information on the physical states of the quantum numbers of interest. The problem of analyzing QCD sum rules hence boils down to extracting the properties of the spectral function from these weighted integrals.

The traditional method for tackling this problem is to introduce a phenomenological ansatz for parametrizing the spectral function, which contains a small number of parameters (most commonly used is the "pole + continuum" ansatz) and then to fit these parameters to the sum rules. This analysis has often worked well in cases, in which one already has some prior information on the form of the spectral function, but is clearly not satisfactory if one can not rely on such prior information. Therefore, it has been argued [1] that the maximum entropy method (MEM) can be used for the analysis of QCD sum rules in order to obtain results, which are less dependent on a-priori assumptions. To verify this claim, the ρ meson channel was studied [1], after which the nucleon channel without and with parity projection [4, 5], and quarkonium spectral functions at both zero and finite temperature [6, 7], have been investigated as well, confirming the usefulness of this novel approach. Recently, the behavior of the ϕ meson in nuclear matter also has been studied [8]. In these proceedings, we will briefly review the obtained results on quarkonium at finite temperature and discuss our findings on the ϕ meson in dense matter.

The results presented in these proceedings have been obtained in collaboration with Kei Suzuki, Keisuke Ohtani, Kenji Morita and Makoto Oka.

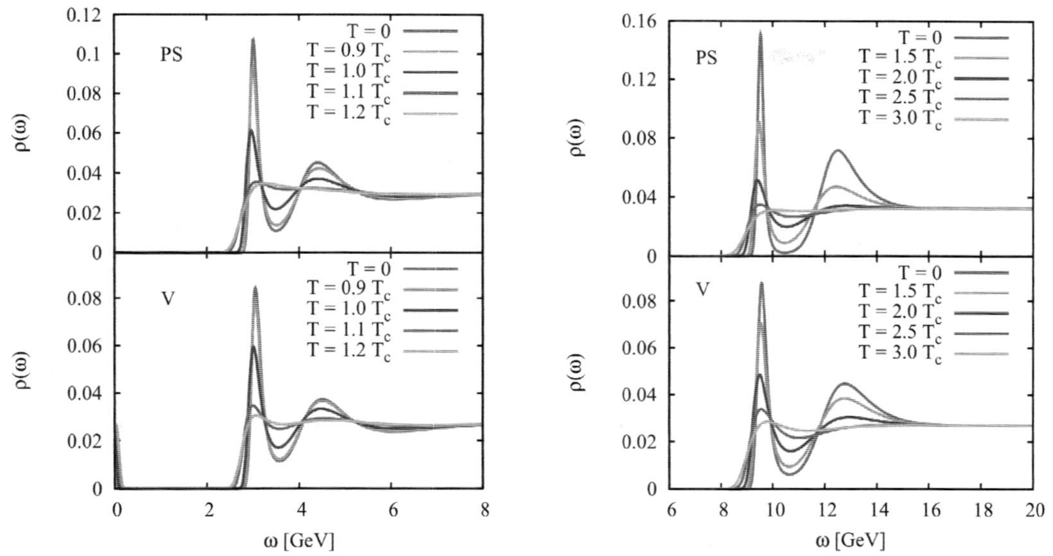

Figure 1. Charmonium (bottomonium) spectral functions at zero and finite temperature in the pseudoscalar (top left (right)) and vector (bottom left (right)) channels. The left (right) figures are adapted from Ref. [6] ([7]).

2. Quarkonia at finite temperature

Let us first discuss the results of the charmonium channels at finite T [6], shown on the left side of Fig. 1. Concentrating firstly on the spectral functions at zero temperature, it is seen that clear peaks are generated, which represent the lowest state of both (pseudoscalar and vector) channels. The positions of these peaks reproduce the experimental values with a precision of about 50 MeV. At finite temperature, we observe that the lowest peaks vanish slightly above the critical temperature T_c. The origin of this melting effect is a sudden change of the gluonic condensates around T_c, which can be related to the deconfinement transition of the gluonic matter.

Next, let us look at the results for bottomonium [7] (see at right side of Fig. 1). Here, as for charmonium, clear peaks are seen for the pseudoscalar and vector channels at zero temperature. These peaks are found at 150-500 MeV above the experimental values of the respective ground states. This discrepancy is caused by the excited states (for instance $\Upsilon(2S)$ and $\Upsilon(3S)$ in the vector channel), which cannot be resolved by the MEM analysis, and pull the lowest peaks to higher energies than the actual ground state. Turning to the finite temperature curves, it is noted that the bottomonium states are modified much slower than their charmonium counterparts, which is in agreement with phenomenological expectations. Concretely, the spectral functions exhibit a clear peak at $T = 2.0\,T_c$ which starts to dissolve at about $2.5\,T_c$.

3. The ϕ meson at finite density

The study of light vector mesons at finite density has in recent years attracted much interest because the vector mesons potentially contain information on the partial restoration of chiral symmetry, which can be measured in experiments [9]. Specifically, the ϕ meson mass is strongly related to the strange quark condensate $\langle \bar{s}s \rangle$, which is modified in matter. This modification is, in the linear density approximation, quantified through the strange sigma term σ_{sN}:

$$\langle \bar{s}s \rangle_\rho = \langle 0|\bar{s}s|0 \rangle + \frac{\sigma_{sN}}{m_s}\rho. \tag{1}$$

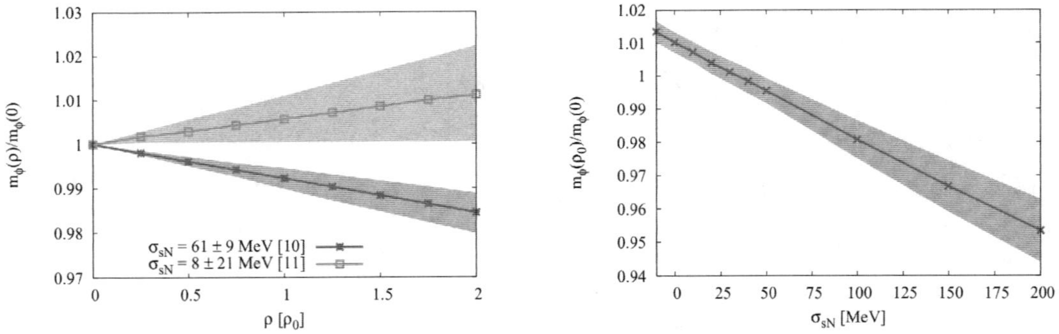

Figure 2. (Left plot) The ϕ meson mass as a function of density for two typical values of the strange sigma term σ_{sN} obtained in recent lattice QCD calculations. The ϕ meson mass is given relative to its vacuum value, while the density is given in units of the nuclear matter density ρ_0. (Right plot) The ϕ meson mass at normal nuclear matter denisty ρ_0, as a function of σ_{sN}.

Studying the mass of the ϕ meson as a function of density within our approach, it is found that the mass shift caused by the finite density effects depends strongly on the value of σ_{sN} [8]. This is shown firstly on the left plot of Fig. 2, where the ϕ meson mass is shown as a function of density for two typical values of σ_{sN} obtained in recent lattice QCD calculations [10, 11]. The dependence becomes even more clear on the right plot of Fig. 2, where the ϕ meson mass at normal nuclear matter density ρ_0 is shown as a function of σ_{sN}. It is seen in this plot that the ϕ meson mass shift at ρ_0 depends linearly on σ_{sN} and is positive for $\sigma_{sN} < 35\,\text{MeV}$, while becoming negative for for $\sigma_{sN} > 35\,\text{MeV}$.

4. Summary

We have in these proceedings discussed two applications of a novel analysis method for QCD sum rules, based on MEM. This method allows us to study the sum rules without making any strong assumptions on the form of the spectral function and therefore provides a more general framework than the traditional analysis methods for QCD sum rules.

We have firstly reviewed our studies on quarkonium at finite temperatures [6, 7], in which we could for the first time describe the melting of both charmonium and bottomonium ground states in a QCD sum rule analysis. Next, we have presented our results on the behavior of the ϕ meson at finite density [8] and have discussed the strong relationship between the mass shift of the ϕ meson at finite density and the strange sigma term $\sigma_{sN} = m_s\langle N|\bar{s}s|N\rangle$.

This work was supported by RIKEN Foreign Postdoctoral Researcher Program and the RIKEN iTHES Project.

[1] P. Gubler and M. Oka, Prog. Thor. Phys. **124**, 995 (2010).
[2] M.A. Shifman, A.I. Vainshtein, and V.I. Zakharov, Nucl. Phys. **B147**, 385 (1979); **B147**, 448 (1979).
[3] L.J. Reinders, H. Rubinstein, and S. Yazaki, Phys. Rep. **127**, 1 (1985).
[4] K. Ohtani, P. Gubler and M. Oka, Eur. Phys. J. A **47**, 114 (2011).
[5] K. Ohtani, P. Gubler and M. Oka, Phys. Rev. D **87**, 034027 (2013).
[6] P. Gubler, K. Morita and M. Oka, Phys. Rev. Lett. **107**, 092003 (2011).
[7] K. Suzuki, P. Gubler, K. Morita and M. Oka, Nucl. Phys. **A897**, 28 (2013).
[8] P. Gubler and K. Ohtani, arXiv:1404.7701 [hep-ph].
[9] T. Hatsuda and S.H. Lee, Phys. Rev. C **46**, 34 (1992).
[10] W. Freeman and D. Toussaint (MILC Collaboration), Phys. Rev. D **88**, 054503 (2013).
[11] H. Ohki *et al.* (JLQCD Collaboration), Phys. Rev. D **87**, 034509 (2013).

4th International Hadron Physics Conference (TROIA'14) IOP Publishing
Journal of Physics: Conference Series **562** (2014) 012012 doi:10.1088/1742-6596/562/1/012012

Status of light scalar mesons as non-ordinary mesons

J.R. Peláez

Departamento de Física Teórica II. Universidad Complutense.28040 Madrid-SPAIN

E-mail: jrpelaez@fis.ucm.es

Abstract. In this talk I briefly review the status of the $f_0(500)$ and $f_0(980)$ together with the other light scalar resonances, as well as the emerging picture of a non-ordinary light meson multiplet, paying particular attention to unitarized Chiral Perturbation Theory, large N_c, semi-local duality and Regge theory arguments.

1. Introduction and present status

The lightest scalar meson, recently renamed $f_0(500)$, but traditionally known as the σ meson, has been a longstanding puzzle in our understanding of strong interactions. This resonance is very relevant both for Nuclear and Particle Physics. For the former because, on a first approximation, it is largely responsible for nucleon-nucleon attraction, and for the latter due to its role in spontaneous chiral symmetry breaking and the identification of glueballs - two fundamental features of QCD. Moreover, there is an ongoing worldwide experimental program on strong interactions (CERN, China, Germany, Japan & USA), even looking for New Physics, where many results depend crucially on final state interactions of pions, producing scalar mesons in general and the $f_0(500)$ in particular. Despite this relevance, the σ properties, nature, and even its very existence have been controversial for decades, since it was first proposed almost 60 years ago. As an example of this debate, the σ (with different names) appeared listed in the early editions of the Particle Data Tables (PDT) [1], disappeared for quite some time of those lists, and made it back to the tables in 2002. Surprisingly, despite being quoted with a huge mass uncertainty from 400 to 1200 MeV and a huge uncertainty in the width, it was nevertheless considered a "well established state".

Finally, and after more than five decades, the controversy on the $f_0(500)$ mass and width seems to be settled. This has been rather well known to specialists for quite some time [2, 4, 5, 6, 3, 7, 8, 9, 10] but remained less well known to the Particle and Nuclear Physics community at large. Actually, this situation has been finally acknowledged with a major revision of its properties in the 2012 edition of the PDT[1], where, after more than a decade of quoting such a disproportionately huge mass uncertainty, it has been reduced by a factor of more than five (and the width uncertainty by more than two), settling the case for the so-called "light sigma" scenario, and opening the door for an even more radical revision in the future, in terms of current dispersive analyses (see the "Note on light scalars" in the PDT).

Of course, the evidence that supports this recent recognition has been accumulating over many years both from the experimental and theoretical side, using a wide variety of techniques, which not only have allowed for precise and model independent determinations of the mass and width of this meson, but which have advanced our understanding of its non-ordinary features.

In addition, in the 2012 PDT edition, the $f_0(980)$ has also suffered a minor revision, by changing the central value of its mass from 980 to 990 MeV whereas its mass uncertainty has doubled to 20 MeV. Other possible multiplet partners of the $f_0(500)$, as the $a_0(980)$ and $K(800)$ have not suffered any changes in their quoted parameters in the last PDT edition. Note that the $K(800)$ is still considered as "not well established" despite its pole has already been well established within a rigorous dispersive approach [36]. The $K(800)$ is therefore in a situation somewhat similar to that of the $f_0(500)$ a few years ago: Its existence is widely accepted by the "scalar-community" but it just lacks recognition in the PDT.

Historically, the first proposal of a scalar isoscalar field was made in the 50's [11] and it was soon realized by J. Schwinger that expectation values of such a scalar singlet field, which he called σ [12], could be of relevance for the breaking of symmetries. Actually it soon became relevant for simple models of the attractive part of the nucleon-nucleon interaction and for the spontaneous chiral symmetry breaking of QCD. For the latter one should mention the traditional Linear Sigma Model [13], but also the more modern and systematic approach of Chiral Perturbation Theory [15], including its unitarization [16, 17, 4, 5, 6, 18], which is needed to generate resonances. Apart from the questioning of their existence, light scalars have faced the controversy on their classification in multiplets [19, 2, 9, 10] leading to the suggestion that they may have a predominantly non-ordinary quark-antiquark nature [19, 20], which has received a recent strong support based on the N_c dependence [21, 22, 23, 24, 25], unusual Regge behavior [26], etc... On the experimental side, the clarification of these controversial issues was hindered by the difficult observation from nucleon-nucleon interactions and the large systematic uncertainties in traditional pion-nucleon experiments used to extract pion-pion scattering [27]. However, more recent "production" experiments, with different systematics from scattering processes, have been very relevant in the recent clarification in the PDT (like [28]) although these also may have problems of their own (use of Breit-Wigner parametrizations, isobar models, etc...). Finally, the results that have partly driven the major σ revision in the PDT have been the latest NA48/2 data on [30] $K \to \pi\pi\ell\nu$ experiments (some older can be found on [29]) together with the most recent dispersive approaches that have provided a rigorous treatment of this resonance.

2. The $f_0(500)$ parameters

As just commented, one of the main difficulties to present a convincing case for the existence of a σ resonance pole around 500 MeV and not higher —the so-called "light sigma" scenario— were due to: i) the poor data on $\pi\pi$ scattering [27], ii) its large width and iii) the fact that the σ does not present the familiar peak and shape of a Breit-Wigner resonance. A description in terms of poles in the complex plane is more appropriate, although, being so wide, its associated pole is deep in the complex plane. This demands a careful, rigorous and model independent analytic extension after the amplitudes that describe the data have been constrained with dispersion relations. The precise analyses of $\pi\pi$ scattering with dispersion relations [38, 33, 40] and their later use to find the associated poles [33, 34, 35, 36] have driven the major revision in the PDT [1], which now reads 400−550 MeV for the mass and 400−700 MeV for the width, as defined from the associated pole. Actually, the PDT mass and with new estimates are "conservative", and the PDT also provides a "more radical point of view", relying only on such model independent dispersive approaches, yielding a mass of 446 ± 6 MeV and a width of 552 ± 10 MeV.

The dispersive results are also responsible for the minor change in the $f_0(980)$ parameters and, when applied to πK scattering, for the most reliable $K(800)$ determination [36]. I have no space to discuss them here in detail but can only refer to the appropriate references on Roy equations [39, 38, 33, 36], GKPY equations [40], Forward Dispersion relations [32, 41] and some pedagogical introductions [31, 32].

3. Chiral Symmetry and the $f_0(500)$

Within Particle Physics, part of the interest on the $f_0(500)$ is due to its role in the spontaneous chiral symmetry breaking of QCD. Actually, the traditional name "sigma", given by J.Schwinger [12], was later adopted by the classic 'Linear Sigma Model" (LSM) formulation of such a breaking [13]. The LSM has played a very important role in our phenomenological and more intuitive description of low energy hadronic interactions, which has its success and shortcomings, leading to multiple variations, i.e. [3]. Similarly, a relatively light scalar-isoscalar meson coupling strongly to $\pi\pi$, and therefore very wide, also seems to be generated in a Nambu-Jona-Lasinio (NJL) model or its modifications within Hadron Physics [14]. Still, both the LSM and NJL are simple models that, despite capturing many relevant features, do not provide a systematic description of low energy hadron Physics and a clear connection to QCD, and usually require some ad-hoc modifications. Nevertheless, a systematic analysis exists under the form of the more modern Chiral Perturbation Theory (ChPT), which is the low energy effective Theory of QCD. Within standard ChPT [15], only Goldstone modes appear explicitly in the Lagrangian, and the dynamics is encoded in a set of "low energy parameters", whose values are saturated [37] by heavier resonances, but not by the sigma, which is another hint of its non-ordinary nature. ChPT is limited to low energies and resonances are not present in its Lagrangian, but can be generated when combined with some dispersive techniques. This approach, less systematic, known as unitarized Chiral Perturbation Theory [16, 17], has been very successful in the description of light scalars and is particularly simple in the case of the $f_0(500)$ [4]. These unitarization techniques allow to study the quark mass dependence of the σ [42], also recently studied on the lattice despite the difficulties in providing results for physical quark masses [43]. Other models in the literature deal with the $f_0(500)$ together with the other light scalars, like extended linear sigma models [3], non-linear models [9, 10], as well as the simple but very successful Chiral Unitary Approach [5, 6], which is tightly connected to Unitarized Chiral Perturbation Theory described before. The lesson to learn from all them is that even very simple descriptions of low energy data, when implementing some basic chiral symmetry constraints and using a decent analytic extrapolation, yield a $f_0(500)$ pole in fair agreement with the precise and model independent dispersive determinations commented in the previous section.

4. The spectroscopy and nature of light scalars

These are the $f_0(500)$ features that are still under some discussion, although the situation has also clarified considerably over the last years. The first problem with spectroscopy is how light scalars are assigned to U(3) multiplets. There is a very strong support [19, 2, 3, 5, 6, 8] for the sigma to belong to a nonet with the other lightest scalars, namely the $f_0(980)$, $a_0(980)$ and $K(800)$. Still, this assignment is obscured by the possible different mixing schemes [45, 10] within states of the multiplet (or even other multiplets) as well as by the controversial status of some scalar mesons, like the $K(800)$ or the $f_0(1370)$.

Somewhat more controversial is the composition of these resonances in terms of quarks and gluons. It was long ago suggested that the ordinary identification of mesons as $q\bar{q}$ states did not match well the light scalars due to their inverted hierarchy [19] as well as their large widths and couplings [44]. Of course, attention has to be paid to what is meant by "composition", since the intuitive Foch space decomposition is not always well defined. Most interpretations are made within some kind of "constituent models". However there are recent attempts to tackle this problem from QCD via the $1/N_c$ behavior of the $f_0(500)$ pole generated from unitarized ChPT [22, 8, 23], or, more recently, also on the lattice [49]. Unitarized ChPT supports a predominantly non-$q\bar{q}$ nature for the $f_0(500)$ providing a very strong support for the $f_0(500)$ meson, since it depends on N_c in a rather different way than predicted from QCD for $q\bar{q}$ mesons [46]. This effect was also observed in $\pi\pi$ scattering by explicit introduction of resonant states [21, 24]. Furthermore, the same conclusion can be obtained without using a particular model, but just from data, the pole positions and the general behavior of $q\bar{q}$ states in QCD [25]. Moreover, larger N_c extrapolations [50] suggest that, even if it is not predominantly an ordinary meson, it might contain some small admixture of a heavier $q\bar{q}$ state [23]. Note this is the $1/N_c$ expansion around $N_c = 3$ [47], although some results are also available for the "large N_c limit" [48].

Finally, since the $f_0(500)$ has not been commonly included in the ordinary classifications of hadrons into linear Regge trajectories with a universal slope [51], its Regge trajectory has been shown not to follow a straight line and that its slope is of a very different order of magnitude than ordinary mesons [26], once again supporting the non-ordinary nature of this state.

5. Acknowledgements

I thank the organizers of Troia14 for their kind hospitality and for creating such a nice scientific environment. Work supported by the Spanish Research contract FPA2011-27853-C02-02 and the EU-Research Infrastructure Integrating Activity "Study of Strongly Interacting Matter" (acronym HadronPhysics2, Grant Agreement n. 227431, 7th Framework Programme).

References

[1] J. Beringer et al. (Particle Data Group), Phys. Rev. D86, 010001 (2012).
[2] J. D. Weinstein and N. Isgur, Phys. Rev. Lett. **48**, 659 (1982). E. van Beveren,*et al.* Z. Phys. C **30**, 615 (1986). U. G. Meissner, Comments Nucl. Part. Phys. **20**, 119 (1991). J. A. Oller and E. Oset, Phys. Rev. D **60** (1999) 074023. E. van Beveren and G. Rupp, Eur. Phys. J. C **22**, 493 (2001). V. V. Anisovich, Phys. Usp. **47**, 45 (2004) [Usp. Fiz. Nauk **47**, 49 (2004)] [hep-ph/0208123]. J. Vijande, A. Valcarce, F. Fernandez and B. Silvestre-Brac, Phys. Rev. D **72**, 034025 (2005). N. N. Achasov and A. V. Kiselev, Phys. Rev. D **73**, 054029 (2006) [Erratum-ibid. D **74**, 059902 (2006)], L. Maiani, A. D. Polosa and V. Riquer, Phys. Lett. B **651**, 129 (2007). G. 't Hooft, *et al.* Phys. Lett. B **662**, 424 (2008). T. Hyodo, D. Jido and T. Kunihiro, Nucl. Phys. A **848**, 341 (2010).
[3] C. Rosenzweig, J. Schechter and C. G. Trahern, Phys. Rev. D **21**, 3388 (1980). J. Schechter and Y. Ueda, Phys. Rev. D **3**, 2874 (1971) [Erratum-ibid. D **8**, 987 (1973)]. R. Delbourgo and M. D. Scadron, Mod. Phys. Lett. A **10**, 251 (1995) [hep-ph/9910242]. D. Black,A. H. Fariborz, F. Sannino and J. Schechter, Phys. Rev. D **59**, 074026 (1999). D. Black, A. H. Fariborz, S. Moussa, S. Nasri and J. Schechter, Phys. Rev. D **64**, 014031 (2001) [hep-ph/0012278]. F. E. Close and N. A. Tornqvist, J. Phys. G **28**, R249 (2002) F. Giacosa, Phys. Rev. D **74**, 014028 (2006). N. N. Achasov and G. N. Shestakov, Phys. Rev. Lett. **99**, 072001 (2007) D. Parganlija, F. Giacosa and D. H. Rischke, Phys. Rev. D **82**, 054024 (2010)
[4] A. Dobado and J. R. Pelaez, Phys. Rev. D **56**, 3057 (1997)
[5] J. A. Oller and E. Oset, Nucl. Phys. A **620**, 438 (1997) [Erratum-ibid. A **652**, 407 (1999)].
[6] J. A. Oller, E. Oset and J. R. Pelaez, Phys. Rev. Lett. **80** (1998) 3452,
[7] N. A. Tornqvist and M. Roos, Phys. Rev. Lett. **76**, 1575 (1996) M. Harada, F. Sannino and J. Schechter, Phys. Rev. D **54**, 1991 (1996) M. Harada, F. Sannino and J. Schechter, Phys. Rev. Lett. **78**, 1603 (1997) Z. Y. Zhou, G. Y. Qin, P. Zhang, Z. Xiao, H. Q. Zheng and N. Wu, JHEP **0502**, 043 (2005). R. Garcia-Martin and J. R. Pelaez, F. J. Yndurain, Phys. Rev. D **76**, 074034 (2007) I. Caprini, Phys. Rev. D **77**, 114019 (2008) Y. Mao, X. G. Wang, O. Zhang, H. Q. Zheng and Z. Y. Zhou, Phys. Rev. D **79**, 116008 (2009) G. Mennessier, S. Narison and X. G. Wang, Phys. Lett. B **688**, 59 (2010).
[8] J. R. Pelaez, Mod. Phys. Lett. A **19**, 2879 (2004)
[9] F. Giacosa, T. .Gutsche, V. E. Lyubovitskij and A. Faessler, Phys. Lett. B **622**, 277 (2005)
[10] F. Giacosa, T. .Gutsche, V. E. Lyubovitskij and A. Faessler, Phys. Rev. D **72**, 094006 (2005)
[11] M. H. Johnson and E. Teller, Phys. Rev. **98**, 783 (1955).
[12] J. S. Schwinger, Annals Phys. **2**, 407 (1957).
[13] M. Gell-Mann and M.Levy, Nuovo Cim. **16**, 705 (1960).
[14] Y. Nambu and G. Jona-Lasinio, Phys. Rev. **122**, 345 (1961). T. Hatsuda and T. Kunihiro, Phys. Rept. **247**, 221 (1994)
[15] S. Weinberg, Physica **A96** (1979) 327. J. Gasser and H. Leutwyler, Annals Phys. **158** (1984) 142; and Nucl. Phys. B **250**, 465 (1985).
[16] T. N. Truong, Phys. Rev. Lett. **61** (1988) 2526. Phys. Rev. Lett. **67**, (1991) 2260; A. Dobado et al., Phys. Lett. **B235** (1990) 134.
[17] A. Dobado and J. R. Pelaez, Phys. Rev. D **47**, 4883 (1993)
[18] A. Gomez Nicola and J. R. Pelaez, Phys. Rev. D **65**, 054009 (2002). F. Guerrero and J. A. Oller, Nucl. Phys. B **537**, 459 (1999) [Erratum-ibid. B **602**, 641 (2001)]

[19] R. L. Jaffe, Phys. Rev. D **15**, 267 (1977).
[20] R. L. Jaffe, AIP Conf. Proc. **964**, 1 (2007) [Prog. Theor. Phys. Suppl. **168**, 127 (2007)] R. L. Jaffe, Nucl. Phys. A **804**, 25 (2008).
[21] F. Sannino and J. Schechter, Phys. Rev. D **52**, 96 (1995)
[22] J. R. Pelaez, Phys. Rev. Lett. **92**, 102001 (2004).
[23] J. R. Pelaez and G. Rios, Phys. Rev. Lett. **97**, 242002 (2006)
[24] J. Nieves, A. Pich and E. Ruiz Arriola, Phys. Rev. D **84**, 096002 (2011).
[25] J. Nebreda, J. R. Pelaez and G. Rios, Phys. Rev. D **84**, 074003 (2011)
[26] J. T. Londergan, J. Nebreda, J. R. Pelaez and A. Szczepaniak, Phys. Lett. B **729**, 9 (2014)
[27] Protopopescu, S. D., et al., *Phys Rev.* **D7**, 1279, (1973). Grayer, G., et al., *Nucl. Phys.* **B75**, 189, (1974). Losty, M. J., et al. *Nucl. Phys.*, **B69**, 185 (1974). Hyams, B., et al., *Nucl. Phys.* **B100**, 205, (1975). P. Estabrooks and A. D. Martin, Nucl. Phys. B **79**, 301 (1974). W. Hoogland *et al.*, Nucl. Phys. B **126**, 109 (1977). M. J. Losty *et al.*, Nucl. Phys. B **69**, 185 (1974). N. B. Durusoy *et al.*, Phys. Lett. B **45**, 517 (1973).
[28] E. M. Aitala *et al.* [E791 Collaboration], Phys. Rev. Lett. **86**, 770 (2001) Phys. Rev. Lett. **89**, 121801 (2002) I. Bediaga and J. M. de Miranda, Phys. Lett. B **633**, 167 (2006) M. Ablikim *et al.* [BES Collaboration], Phys. Lett. B **598**, 149 (2004) D. V. Bugg, Phys. Lett. B **572**, 1 (2003) [Erratum-ibid. B **595**, 556 (2004)], Phys. Rept. **397**, 257 (2004).
[29] L. Rosselet, *et al.*, Phys. Rev. **D15**, 574 (1977). S. Pislak *et al.* [BNL-E865 Collaboration], Phys. Rev. Lett. **87** (2001) 221801.
[30] J. R. Batley *et al.* [NA48-2 Collaboration], Eur. Phys. J. **C70**, 635-657 (2010).
[31] A. D. Martin, T. D. Spearman, *Elementary Particle Theory* North-Holland Pub. Co., 1970. B.R.Martin, D. Morgan and G. Shaw, "Pion-pion Interactions in Particle Physics", Academic Press 1976.
[32] F. J. Yndurain, hep-ph/0212282.
[33] G. Colangelo, J. Gasser and H. Leutwyler, Nucl. Phys. B **603**, 125 (2001).
[34] I. Caprini, G. Colangelo and H. Leutwyler, Phys. Rev. Lett. **96**, 132001 (2006).
[35] R. Garcia-Martin, R. Kaminski, J. R. Pelaez and J. Ruiz de Elvira, Phys. Rev. Lett. **107**, 072001 (2011).
[36] B. Moussallam, Eur. Phys. J. C **71**, 1814 (2011).
[37] G. Ecker, J. Gasser, A. Pich, E. de Rafael, Nucl. Phys. **B321**, 311 (1989). J. F. Donoghue, C. Ramirez, G. Valencia, Phys. Rev. **D39** (1989) 1947.
[38] B. Ananthanarayan, G. Colangelo, J. Gasser and H. Leutwyler, Phys. Rept. **353**, 207 (2001).
[39] S. M. Roy, Phys. Lett. B **36**, 353 (1971). M. R. Pennington and S. D. Protopopescu, Phys. Rev. D **7** (1973) 1429 and Phys. Rev. D **7**, 2591 (1973). J. L. Basdevant, C. D. Froggatt and J. L. Petersen, Phys. Lett. B **41** (1972) 173; Phys. Lett. B **41** (1972) 178. Nucl. Phys. B **72**, 413 (1974). S. Descotes-Genon, N. H. Fuchs, L. Girlanda and J. Stern, Eur. Phys. J. C **24**, 469 (2002) R. Kaminski, L. Lesniak and B. Loiseau, Phys. Lett. B **551**, 241 (2003) R. Kaminski, J. R. Pelaez and F. J. Yndurain, Phys. Rev. D **77**, 054015 (2008)
[40] R. Garcia-Martin,*et al.* Phys. Rev. **D83**, 074004 (2011).
[41] J. R. Pelaez and F. J. Yndurain, Phys. Rev. D **71**, 074016 (2005) [hep-ph/0411334].
[42] C. Hanhart, J. R. Pelaez and G. Rios, Phys. Rev. Lett. **100**, 152001 (2008) J. Nebreda and J. R. Pelaez., Phys. Rev. D **81**, 054035 (2010)
[43] T. Kunihiro, S. Muroya, A. Nakamura, C. Nonaka, M. Sekiguchi and H. Wada, Nucl. Phys. Proc. Suppl. **186**, 294 (2009) K. Jansen *et al.* [ETM Collaboration], Phys. Rev. D **80**, 054510 (2009) S. Prelovsek, T. Draper, C. B. Lang, M. Limmer, K. -F. Liu, N. Mathur and D. Mohler, Phys. Rev. D **82**, 094507 (2010) T. Kunihiro, S. Muroya, A. Nakamura, C. Nonaka, M. Sekiguchi and H. Wada, EPJ Web Conf. **3**, 03010 (2010). J. Soto, P. Talavera and J. Tarrus, Nucl. Phys. B **866**, 270 (2013) [arXiv:1110.6156 [hep-ph]].
[44] M. R. Pennington, Phys. Rev. Lett. **97**, 011601 (2006). J. A. Oller, L. Roca, Eur. Phys. J. **A37**, 15-32 (2008). J. Bernabeu, J. Prades, Phys. Rev. Lett. **100**, 241804 (2008).
[45] J. A. Oller, Nucl. Phys. A **727**, 353 (2003) A. H. Fariborz, R. Jora and J. Schechter, Phys. Rev. D **72**, 034001 (2005) [hep-ph/0506170].
[46] G. 't Hooft, Nucl. Phys. B **72** (1974) 461. E. Witten, Annals Phys. **128** (1980) 363.
[47] J. R. Pelaez, J. Nebreda and G. Rios, Prog. Theor. Phys. Suppl. **186**, 113 (2010) J. R. Pelaez, arXiv:hep-ph/0509284 (11th International Conference on Elastic and Diffractive Scattering, Blois, France, 2005). J. R. Pelaez and G. Rios, arXiv:0905.4689 [hep-ph]. (Excited QCD, Zakopane, Poland,2009).
[48] J. Nieves and E. Ruiz Arriola, Phys. Lett. B **679**, 449 (2009) J. Nieves and E. Ruiz Arriola, Phys. Rev. D **80**, 045023 (2009) Z. X. Sun, *et al.*hep-ph/0411375 and hep-ph/0503195.
[49] G. S. Bali, *et al.* JHEP **1306**, 071 (2013)
[50] J. Ruiz de Elvira, J. R. Pelaez, M. R. Pennington and D. J. Wilson, Phys. Rev. D **84**, 096006 (2011)
[51] A. V. Anisovich, V. V. Anisovich and A. V. Sarantsev, Phys. Rev. D **62**, 051502 (2000).

Observation of Structures in the $J/\psi\phi$ Mass Spectrum at CMS

E. Asli Yetkin [1]

Physics Department, Mimar Sinan Fine Arts University, TURKEY

E-mail: eaa@cern.ch

Abstract. Observation of charmonium like new states since 2003, which do not fit into conventional quark model renewed the interest in exotic states. Many models like meson molecule, tetraquark, and quark-gluon hybrid are proposed to explain the observed charmonium spectroscopy. The observation of Y(3940)[1] near the $J/\psi\omega$ threshold motivated the searches for similar structures near $J/\psi\phi$ threshold. In this proceeding, we summarize the recent developments on the observation of a peaking structure in the $J/\psi\phi$ mass spectrum from $B^{\pm} \to J/\psi\phi K^{\pm}$ decay.

1. Introduction

In 2009, the CDF collaboration has reported an evidence for a narrow structure near the $J/\psi\phi$ threshold in the exclusive $B^{\pm} \to J/\psi\phi K^{\pm}$ decay and called it Y(4140)[2]. In 2011, they updated their results with a larger data set and observed Y(4140) with a significance greater than 5σ[3]. They also observed a second structure in the same mass spectrum, Y(4274), with a significance above 3σ. Y(4140) could be a good candidate for the exotic mesons since it contains both a $s\bar{s}$ and a $c\bar{c}$ pairs, has mass well beyond open charm pair and its width is relatively narrow. Potential explanations for the observed structures include a $D_s^{*+}D_s^{*-}$ molecule, an exotic charmonium hybrid, and a tetraquark state.

After LHC became operational the LHCb collaboration also searched for the Y(4140) structure in $B^{\pm} \to J/\psi\phi K^{\pm}$ decays using the LHC data and reported a result in conflict with CDF result[4]. In order to solve this puzzle, CMS performed an analysis using $5.2 fb^{-1}$ pp collision data collected at LHC at 7 TeV center of mass energy[6]. The $B^{\pm}$ candidates are reconstructed in the decay channel $B^{\pm} \to J/\psi\phi K^{\pm}$, where $J/\psi \to \mu^+\mu^-$ and $\phi \to K^+K^-$. Recently similar search is also reported by D0 collaboration[7]. In this proceeding, details of CMS analysis will be summarized and the results will be compared to the CDF, LHCb, and D0 measurements.

2. Reconstruction of B Signal

In CMS, reconstruction of the B signal starts at data collection level where the data sample is selected on the basis of the dimuon decay mode of J/ψ. The events containing J/ψ candidates are selected by the HLT dimuon trigger and the J/ψ candidates are used in the B reconstruction if the reconstructed dimuon invariant mass is within the 150 MeV of the J/ψ mass[5]. The

[1] On behalf of the CMS Collaboration

$B^{\pm} \to J/\psi\phi K^{\pm}$ candidates are reconstructed by combining additional three charged tracks (total charge of ±1) to the J/ψ vertex. Kaon mass is assigned to the three charged tracks. There are two K^+K^- combinations from the three charged kaon tracks, and the one with a lower invariant mass which satisfy $1.008 < m(K^+K^-) < 1.035$ is accepted as the ϕ candidate. The study of MC simulations of the $B^{\pm}$ decay and data show that the ϕ signal from the other combination is negligible.

The invariant-mass spectrum of the selected $J/\psi\phi K^{\pm}$ candidates is shown in the left plot of Fig.1 for a mass difference $\Delta m < 1.568$GeV where Δm defined as $m(\mu^+\mu^-K^+K^-) - m(\mu^+\mu^-)$. This selection is applied to eliminate the $B_s^0 \to \psi(2S)\phi \to J/\psi\pi^+\pi^-\phi$ background at higher values of Δm.

The invariant-mass spectrum is fit with a Gaussian signal function and a second-degree polynomial background function. The returned mass of 5.2796 ± 0.0006 (stat.) GeV is consistent with the nominal value[5] and the extracted width value of 9.6 ± 0.7 (stat.) MeV agrees with MC prediction. With 2480 ± 160 (stat.) B yield the selected data set is the world's largest $B^{\pm} \to J/\psi\phi K^{\pm}$ sample up to date.

3. Observation of Peaks in the $J/\psi\phi$ Mass Spectrum

To search for possible structures in the $J/\psi\phi$ mass spectrum, the dataset is separated in to the 20 MeV-wide Δm intervals from 1.008 to 1.568 GeV and the $J/\psi\phi K^{\pm}$ mass distributions for each bin are fit to extract the B signal yield in that Δm interval. A double Gaussian function is used for the signal and a 2nd degree polynomial for the combinatorial background. The mean values of the two Gaussian signals are fixed to the nominal B mass, the with of the Gaussians and the relative fractions are fixed to the values obtained from the signal simulation for each Δm bin. The obtained Δm spectrum is after correcting for relative efficiency is shown by black dot marker in Fig.2.

The mass and the widths values of the observed two structures are extracted by fitting the $J/\psi\phi K^{\pm}$ invariant-mass distribution in each Δm interval with a global unbinned maximum-likelihood (UML) fit. In each fit interval, the $B^{\pm}$ mass is fixed to its nominal value and the mass resolution is fixed to the value calculated from simulation. The combinatorial background in each fit is modeled as a second-degree polynomial. In the global fit, the $B^{\pm}$ yield is expressed

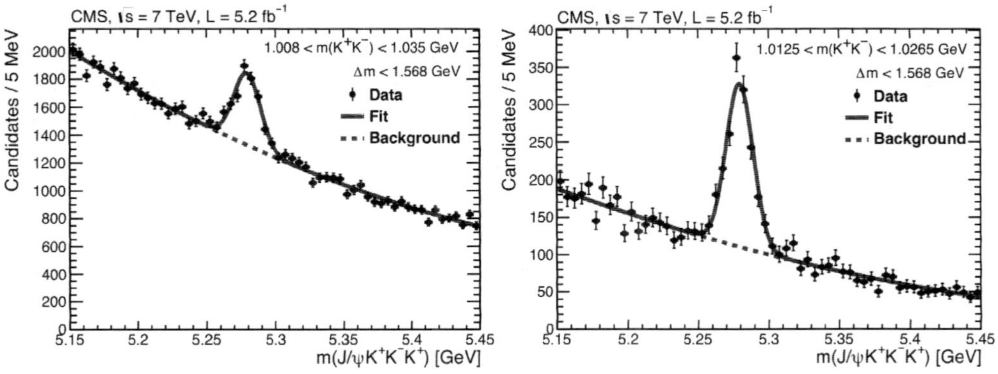

Figure 1. The $J/\psi\phi K^{\pm}$ mass distribution with the standard event selection (left) and the tighter requirements (right). The solid curves show the result of fitting these distributions to a Gaussian signal and a second-degree polynomial background while the dashed curves show the background contribution[6].

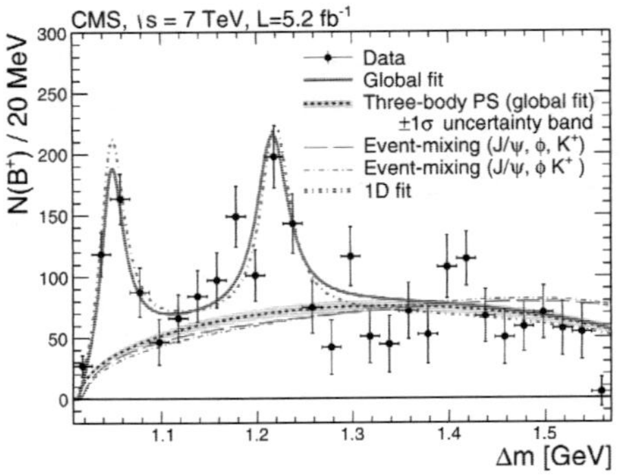

Figure 2. The number of $J/\psi\phi K^{\pm}$ candidates as a function of $\Delta m = m(\mu^+\mu^-K^+K^-) - m(\mu^+\mu^-)$. The solid curve is the global unbinned maximum-likelihood fit of the data, and the dotted curve is the background contribution assuming three-body PS. The band is the $\pm 1\sigma$ uncertainty range for the background obtained from the global fit. The dashed and dash-dotted curves are background curves obtained from two different event-mixing procedures, as described in the text, and normalized to the number of three-body PS background events. The short dashed curve is the 1D fit to the data[6].

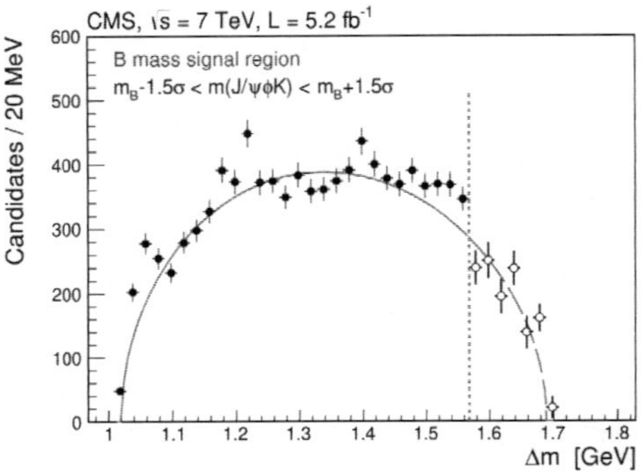

Figure 3. The Δm spectrum, including non-B candidates after subtracting B_s^0 contribution, within $\pm 1.5\sigma$ of the B nominal mass. The dashed vertical line indicates the boundary of the region eliminated from the analysis (right region). The data in the left region are represented by black point markers. The entries in the right region are subtracted of the B_s^0 background component and are represented by open point markers. The red curve is a three-body phase-space distribution, normalized in the left region (solid) after subtracting the yields from the corresponding two structures, and extrapolated to the right region (dashed)[6].

as the product of the relative efficiency times the number of signal events from the two BWs and the nonresonant continuum events. The invariant $J/\psi\phi K^{\pm}$ mass distribution is fit for each Δm bin simultaneously by projecting the above product into each bin. The returned signal yields from the UML fit for each mass structures are 310 ± 70 (stat.) and 418 ± 170 (stat.). The mass and the width values of the first structure are $\Delta m_1 = 1051.3 \pm 2.4$ (stat.) MeV, $\Gamma_1 = 28^{+15}_{-11}$ (stat.) MeV and for the second structure are $\Delta m_2 = 1217.1 \pm 5.3$ (stat.) MeV, $\Gamma_2 = 38^{+30}_{-11}$ (stat.) MeV. The projection of the UML fit assuming two structures onto the $J/\psi\phi$ mass spectrum is represented as the solid line in Fig.2. To check the robustness of the fitting procedure an alternative one-dimensional (1D) binned χ^2 fit is applied to the Δm spectrum. The signal and the background functions are not changed in the 1D fit. The result of the 1D fit, assuming two structures, is shown as the dashed line in Fig.2. The extracted mass, widths, and signal yields from the 1D fit are consistent with the values obtained from the global fit.

The Δm background shape is also investigated with event-mixing technique where Δm spectrum is reconstructed from data (a) by combining J/ψ, ϕ, and $K^{\pm}$ coming from different events (b) by combining J/ψ from one event and ϕ,$K^{\pm}$ from another event. The resulting Δm distribution with two-signal-hypothesis fit is shown in Fig.2, as shown the three-body phase space background which is used as the default background model for this analysis lies above the event-mixing background in all Δm bins in the range of interest.

CMS performed various test to examine the robustness of the observed two structures. Each event selection criteria is individually varied and no bias in the selection procedure is observed. The relative efficiency used to correct the extracted spectrum is varied and the fit is

Table 1. Systematic uncertainties in the measured masses and widths of the two peaking structures.

	$m_1(MeV)$	$\Gamma_1(MeV)$	$m_2(MeV)$	$\Gamma_2(MeV)$
$B^{\pm}$ Background PDF	0.8	7.4	2.6	9.9
$B^{\pm}$ Signal PDF	0.2	3.6	2.7	0.2
Relative efficiency	4.8	6.0	0.9	10.0
Δm binning	3.7	1.5	2.7	0.2
Δm structure PDF	0.8	9.3	0.6	4.9
Δm mass resolution	0.8	6.4	0.6	4.6
Δm background shape	0.2	7.0	0.3	0.2
Selection requirements	0.8	7.8	5.5	1.8
Total	6.3	19	7.3	16

repeated. The result confirm the robustness of the significance of the first structure. The Δm spectrum obtained through a different background subtraction technique (sPlot formalism) is compared to the Δm spectrum shown in Fig.2. No indication of bias is found. The analysis is repeated by requiring a tighter B selection to reduce the combinatorial background. Tighter event selection reduce the background by a factor of ten and keep the 40% of the B signal as shown in Fig.1(right). The extracted Δm spectrum for these events looks similar to Fig.2, where the fitted mass and width values of the two peaking structures are consistent with the results obtained from the nominal data sample.

The Δm distribution for the events that are eliminated to remove the effect of B_s^0 background is also studied. This study is performed to make sure the eliminated events do not cause reflections in the low Δm region. The study show that, after subtraction of the B_s^0 background the Δm distribution of events with $\Delta m >$1.568 GeV is consistent with the prediction based on the three-body phase-space hypothesis for the non-resonant background. The Δm spectrum, including non-B candidates after subtracting B_s^0 contribution is shown in Fig.3.

The estimation of the systematic uncertainties in the mass and the width measurements of the observed two structures are determined by performing several studies on the data and also on the MC simulation. The uncertainties related to the probability density functions (PDFs) for signal and background are studied by using different PDFs such as first, second, and third order polynomials, exponential functions, single and double Gaussian functions. The uncertainty due to the relative efficiency is studied by varying the relative efficiency in various bins and comparing with the 2D efficiencies. Estimating contribution due to the binning of the Δm spectrum is studied by using 10 MeV bins instead of 20 MeV bins. To estimate the uncertainty from the signal fitting function, the fit to the Δm distribution is repeated using either a nonrelativistic BW or a P-wave relativistic BW function for each structure. By varying mass resolution obtained from simulation the uncertainty due to the Δm mass resolution is studied. The overall systematic uncertainties in the measurement of the masses and widths of the two structures are found by adding in quadrature the individual items which are summarized in Table1.

4. Summary

CMS observed a peaking structure in the $J/\psi\phi$ mass spectrum from $B^{\pm} \to J/\psi\phi K^{\pm}$ decays by analyzing the data collected at $\sqrt{s} = 7$ TeV pp collisions at the LHC. A statistical significance of greater than 5σ is found assuming an S-wave relativistic BW lineshape for this structure above a

Table 2. Summary of the mass and width measurements of the two structures observed in the $J/\psi\phi$ mass spectrum.

Experiment	$m_1(MeV)$	$\Gamma_1(MeV)$	$m_2(MeV)$	$\Gamma_2(MeV)$
CDF	$4143.0^{+2.9}_{-3.0} \pm 0.6$	$15.3^{+10.4}_{-6.1} \pm 2.5$	$4274.4^{+8.4}_{-6.7} \pm 1.9$	32.3+21.9±7.6
CMS	4148.0±2.4±6.3	28^{+15}_{-11}±19	4313.8±5.3±7.3	38^{+30}_{-11}±16
D0	4159.0±4.3±6.6	$19.9\pm12.6^{+1.0}_{-8.0}$	4328.5±12.0	30(constrained)
LHCb	no observation		no observation	

three-body PS shape for the non-resonant background. Adding the J/ψ mass [5] to the extracted Δm values, the mass and width are measured to be $m_1 = 4148.0 \pm 2.4(stat.) \pm 6.3(syst.)$ MeV and $\Gamma_1 = 28^{+15}_{-11}$(stat.)±19(syst.) MeV. The measured mass and width are consistent with the Y(4140) values reported by CDF experiment[2]. The recent measurement performed by D0 can also accommodate the first structure[7].

In addition, evidence for a second peaking structure is found in the same mass spectrum, with measured mass and width values of $m_2 = 4313.8 \pm 5.3(stat.) \pm 7.3(syst.)$ MeV and $\Gamma_2 = 38^{+30}_{-11}$(stat.)±16(syst.) MeV. The significance of the second structure could not able to reliably determined due to possible reflections of two body decays. A summary of the mass and the width measurements of CMS and other experiments are summarized in Table2. The two structures are well above the threshold of open charm (DD) decays and have relatively narrow widths. Conventional charmonium mesons with these masses would be expected to have larger widths and to decay predominantly into open charm pairs with small branching fractions into $J/\psi\phi$. Angular analysis of the final state particles for $B^{\pm} \rightarrow J/\psi\phi K^{\pm}$ decay and determination of their J^{PC} can clarify the nature of these structures.

Acknowledgments
Work of author E.Asli Yetkin in Mimar Sinan Fine Arts University is supported by TUBITAK (Bideb-2232) research grant.

References

[1] Belle Collaboration 2005 *Phys. Rev. Lett.* **94** 182002
[2] CDF Collaboration 2009 *Phys. Rev. Lett.* **102** 242002
[3] CDF Collaboration 2011 arXiv:1101.6058 [hep-ex]
[4] LHCb Collaboration 2012 *Phys. Rev.* D **85** 091103
[5] Beringer J et al 2012 *Phys. Rev.* D **86** 010001
[6] CMS Colloboration 2014 *Phys. Lett.* B **734** 261
[7] D0 Collaboration 2014 *Phys. Rev.* D **89** 012004

Journal of Physics: Conference Series **562** (2014) 012014 doi:10.1088/1742-6596/562/1/012014

Production of Hadronic Resonances at LHC Energies with the ALICE Experiment

A. Karasu Uysal for the ALICE Collaboration

KTO Karatay University, Konya, Turkey

E-mail: akarasu@cern.ch

Abstract. Hadronic resonances are strongly decaying particles with extremely short lifetimes of about few *fm/c*, which are comparable to or smaller than the lifetime of the system formed in relativistic collisions. Due to their short lifetimes, resonance particles can be used to investigate the freeze-out mechanisms after hadronization. This contribution summarizes the results of resonance particle productions measured by the ALICE collaboration at LHC from various colliding systems and energies. Measured mass, width, transverse momentum and particle ratios of those resonances have been reported. Theoretical predictions from thermal model calculations are also presented.

1. Introduction

The production of resonances in relativistic collisions is sensitive to the properties of strongly interacting matter produced in these collisions. Modification of the production rates and the in-medium properties of hadronic resonances can be used as signatures of a possible phase transition of nuclear matter to a deconfined state of quarks and gluons [1].

The lifetime of resonances is expected to be less than the lifetime of the formed system. Therefore, resonances are expected to decay, re-scatter and regenerate all the way throughout the kinetic freeze-out [2]. The measurement of the resonance yield can provide information about hadronization and the time evolution of source from chemical to kinetic freeze-out [3].

ALICE [4] is the only LHC detector mainly dedicated to the study of ultrarelativistic heavy-ion collisions, in order to investigate the properties of the Quark-Gluon Plasma (QGP). It is a very well suited device for resonance measurement as it provides extensive particle tracking and identification at mid-rapidity ($|\eta| < 0.9$).

This proceedings present the measured results of the $K^*(892)^0$ and $\phi(1020)$ resonances in pp collisions at $\sqrt{s}$ = 7 TeV, in p–Pb collisions at $\sqrt{s_{NN}}$ = 2.76 TeV and in Pb–Pb collisions at $\sqrt{s_{NN}}$ = 2.76 TeV using the ALICE Detector at the LHC.

2. Reconstruction of Resonances

The Inner Tracking System (ITS), the Time Projection Chamber (TPC), the Time-of-Flight Detector (TOF) and the Vertex-0 (V0) Detector are sub-detectors of ALICE used to measure the resonance production through the detection of its hadronic decay products.

Resonances are reconstructed using their decay channels: $K^* \rightarrow K^+\pi^-$ and $\phi \rightarrow K^+K^-$ by invariant mass method only after kinetic freeze-out, when decay products do not re-scatter. Event-mixing technique [5] is used to remove the combinatorial background present in the invariant mass spectra. After the

subtraction of the combinatorial background, the signal fitted with Breit-Wigner and polynomial functions to describe the peak and the remaining residual background, respectively.
Measured mass and width of K^* and ϕ resonances in Pb–Pb collisions are presented in Figure 1. The mass and width of resonances are in agreement with vacuum (Particle Data Group - PDG [6]) values.

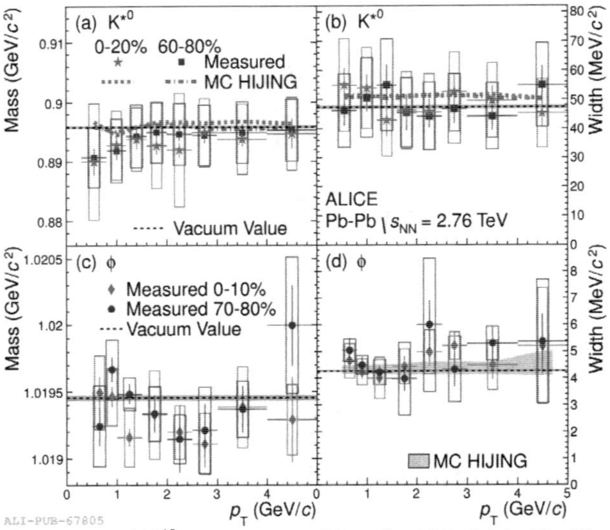

Figure 1. (Upper Panel) Measured K^{*0} meson mass (a) and width (b) in Pb–Pb collisions at 2.76 TeV in the 0-20% and 60-80% centrality intervals, along with the values extracted from Monte-Carlo HIJING simulations. (Lower Panel) Measured ϕ meson mass (c) and width (d) in Pb–Pb collisions at 2.76 TeV in the 0-10% and 70-80% centrality intervals. The ϕ width extracted from HIJING simulations is also shown. The vacuum values of the K^{*0} and ϕ mass and width [PDG] are indicated by the horizontal dashed lines. The statistical uncertainties are shown as bars and the total systematic uncertainties are shown as boxes.

3. Results

The raw counts of K^* and ϕ extracted from the invariant mass fits have been corrected for reconstruction efficiency, acceptance and branching ratio in order to obtain the transverse momentum spectra. Figure 2 shows the corrected transverse momentum spectra of K^* and ϕ spectra in different collision centrality classes in p–Pb collisions at 5.02 TeV. The results on K^* and ϕ production in pp and in Pb–Pb collisions can be found in [7] and [8].

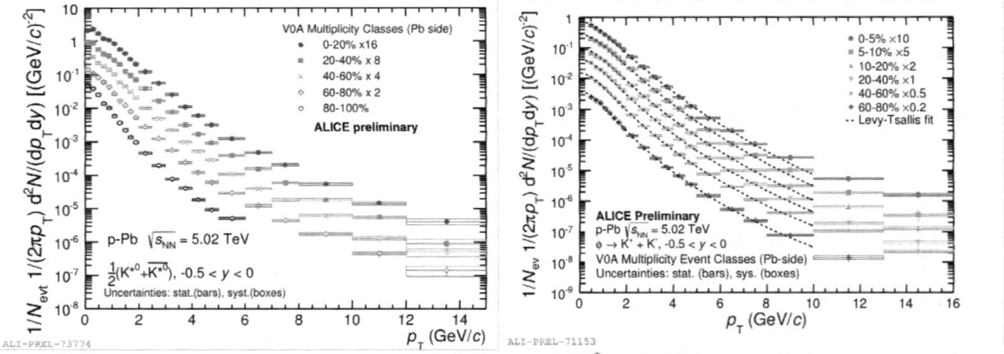

Figure 2. Transverse momentum spectra of (left panel) K^* and (right panel) ϕ resonances in p–Pb collisions at $\sqrt{s_{NN}}$ = 5.02 TeV. Bars and boxes represent statistical and systematic uncertainties, respectively [9].

p/φ, φ/K and K*/K particle ratios presented in Figure 3. In the most central Pb–Pb collisions p/φ ratio is flat below 3 GeV/c. In central p–Pb collisions, the ratio shows a flattening for p_T<1.5 GeV/c and it is similar to 60-80% peripheral Pb–Pb collisions. In peripheral Pb–Pb, peripheral p–Pb and pp collisions the p/φ ratios are similar and decrease rapidly with transverse momentum. The measured ratios φ/K and K*/K presented in right panel of Figure 3. The φ/K ratio is independent from system size and reaches the value predicted by a grand-canonical thermal model with T = 156 MeV [10], the K*/K ratio is decreasing towards more central Pb–Pb collisions, where the measured ratio is about 60 % of the thermal model value.

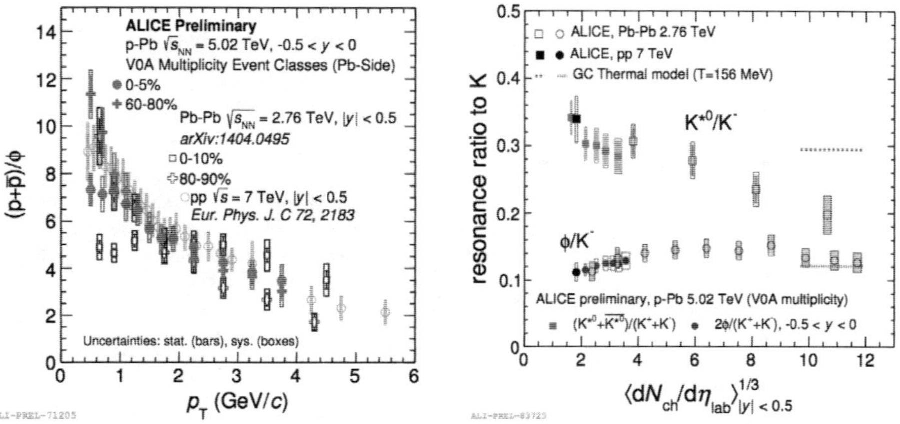

Figure 3. (Left panel) (p+p¯)/φ ratio measured in p–Pb in 0-5% (red dot) and 60-80% (purple cross) V0A multiplicity classes, compared to pp (pink), 0-5% (black hollow squares) and 60-80% (black hollow cross) Pb–Pb collisions. (Right panel) ratio of resonances to charged kaons measured in the three collision systems, as a function of the system size [9].

4. Conclusions

The spectra, mass, and width of the K* and φ mesons have been measured in pp, p–Pb and Pb–Pb collisions using the ALICE detector at LHC energies. Mass and width of the resonances are consistent with the nominal values. Because of the measured K*/K ratio is significantly lower than the thermal model estimation suggests that in central Pb–Pb collisions, K* suffer from re-scattering due to the short lifetime [7].

Acknowledgment

This work was supported by Scientific and Technological Research Council of Turkey (TUBITAK, Project No 113F221).

References

[1] Rapp R and Wambach J 2000 *Adv. Nucl. Phys.* **25** 1
[2] Torrieri G and Rafelski 2001 J *Phys. Lett.* B **509** 239
[3] Bleicher M and Aichelin J, 2002 *Phys. Lett. B* **530** 81
[4] Aamodt K et al (ALICE Collaboration) 2008 *JINST* **3** S08001
[5] L'Hôte D 1994 *Nucl. Inst. Meth. in Phys. Res. Sec.* A **337** 544
[6] Nakamura K et al (Particle Data Group) 2010 *J. Phys. G* **37** 075021
[7] Abelev B et al (ALICECollaboration) 2012 *Eur. Phys. J C* **72** 2183
[8] Abelev B et al (ALICECollaboration) 2014, *preprint* hep-ex/1404.0495
[9] Bellini F 2014 *Nuc. Phys. A*, **08** 31
[10] Stachel J et al 2014 *J. Phys. Conf. Ser.* **509** 012019

Journal of Physics: Conference Series **562** (2014) 012015 doi:10.1088/1742-6596/562/1/012015

Hadronic results from CMS experiment at the LHC

Deniz Sunar Cerci on behalf of the CMS Collaboration

Adiyaman University, Science and Letters Faculty, Physics Department, 02040 Adiyaman, TURKEY

E-mail: deniz.sunar.cerci@cern.ch

Abstract. The Compact Muon Solenoid (CMS) is one of the two large multipurpose experiments at the LHC. We present the CMS results related to tests of QCD predictions for jet, dijet, and multijet production in pp collisions at center-of-mass energies of 7 TeV and 8 TeV. The results also include recent hadronic event shape measurements, α_s extraction, charged hadron measurements, dijet azimuthal decorrelations, searches for BFKL effects in the azimuthal correlations of forward-backward and forward-central jets at 7 TeV, and studies of the transition from the perturbative to the non-perturbative regime using minijets at 8 TeV.

1. Introduction

Events with collimated streams of particles, called jets, are abundantly produced in high energetic proton-proton collisions at the LHC. Such events can be described by Quantum Chromodynamics (QCD) in terms of parton-parton scattering where the outgoing scattered partons manifest themselves as hadronic jets. The benchmarks of the Standard Model are jet and multijet production cross sections which confront the perturbative QCD (pQCD) predictions at the smallest distance scales ever probed. CMS provides unexplored kinematic regimes to test the QCD predictions. A detailed description of the CMS experiment can be found in Ref. [1]. The hadronization and parton shower models can be optimized by using the jet observables. The parton density functions (PDF) can also be constrained with the measurement of jet cross sections which are sensitive to the strong coupling constant α_s. The observables sensitive to multijet production, such as the hadronic event shapes, the dijet azimuthal decorrelations, and the ratio of the 3–jet to 2–jet production cross section, are used to compare data to the predictions of various QCD Monte Carlo (MC) generators.

2. Measurements

Inclusive jet and dijet production cross section measurements are the tools to investigate the structure of the proton. Such a measurement [2] is performed by using the data corresponding to 5 fb^{-1} of integrated luminosity. Jets are reconstructed with the anti-k_T clustering algorithm with distance parameter $R = 0.7$. The measured cross sections are compared to pQCD predictions at NLO, using five sets of PDFs.

The charged particle production, multi-particle production (MPI) and underlying event (UE) activity measurements are important topics to test the soft QCD and used to tune the MC generator parameters. The characteristic features and relative importance of different mechanisms of MPI are studied in pp collisions at $\sqrt{s}$ = 7 TeV [3]. In this study the UE and jets are analyzed for the first time with the observables such as jet p_T spectra, rates and shapes. The results are compared to MC predictions which fail to fully describe the intrajet spectra. In low-multiplicity events, jets are narrower than the MC models, whereas in high-multiplicity events their widths are as predicted by the MC event generators. Most of the MC models reproduce well the UE features in all the multiplicity bins considered.

The inclusive forward jet production cross sections as well as the dijets events with at least one jet emitted at central and the other in forward region are measured at $\sqrt{s} = 7$ TeV in pp collisions [4]. The differential cross sections binned in p_T and η are compared to NLO predictions, MC event generators and models. Within the theoretical and experimental uncertainties, all the predictions are in agreement with the measurement.

In the inclusive jet cross section measurement at $\sqrt{s} = 8$ TeV [5], the data are compared to various pQCD predictions at NLO and found to have good agreement in the entire phase space. Figure 1 shows the comparison of differential inclusive jet cross sections measured at low and high transverse momenta to NLO predictions using the NNPDF2.1 PDF set times the NP correction factor (left) and the ratio of data to the theoretical prediction of NNPDF2.1 (right). The recent inclusive jet cross section measurements at $\sqrt{s} = 8$ TeV are also performed by CMS [6, 7].

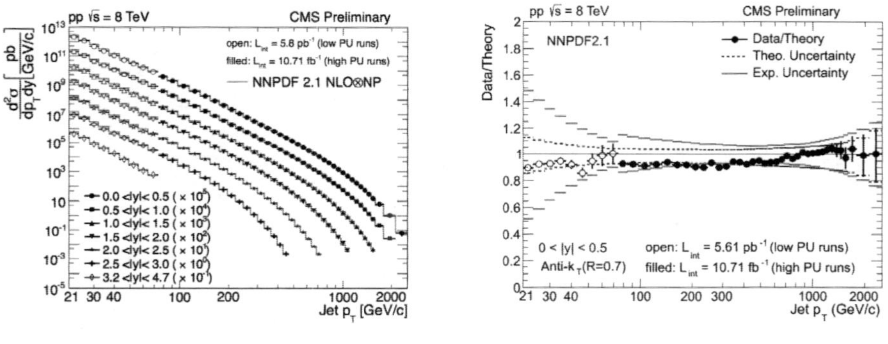

Figure 1. Differential inclusive jet cross sections in comparison to NLO predictions (left). Ratio of data to the prediction of the NNPDF2.1 PDF set (right).

The ratios of 3–jet to 2–jet cross section measurements are done using jets within the central region at $|y| < 2.5$ [8, 9]. In the most recent measurement [10], the value of $\alpha_s(M_Z)$ at the scale of the Z boson mass is measured for the first time at high momentum scales. The result is found as $\alpha_S(M_Z) = 0.1185 \pm 0.0019$ (exp.) ± 0.0028 (PDF) ± 0.0004 (NP)$^{+0.0055}_{-0.0022}$ (scale) which is in agreement with the world average value. The running of $\alpha_s(Q)$ and its total uncertainty as a function of the momentum transfer $Q = p_T$ are shown in Figure 2 (left). An overview of the gluon, sea, u valence (u_V), and d valence (d_V) PDFs at starting scale $Q^2 = 1.9$ GeV^2 is shown in Figure 2 (right). A significant improvement of precision is observed in the high-x region by including the CMS data with DIS data at HERA. The CMS jet data have a larger gluon PDF at high-x compared to the DIS data.

The azimuthal angular decorrelations of jets most forward and backward in rapidity are measured with the CMS detector at 7 TeV [11]. The most forward and the most backward jet pairs in a dijet event are called Muller-Navelet (MN) jet. Due to the highest rapidity separation between MN jets which can be populated with soft emissions, correlation between the jets is an observable which is sensitive to the details of the QCD evolution. BFKL effects can be disentangled from the standard DGLAP approach.

Azimuthal correlations between forward-central jet pairs are also studied with a large separation in pseudorapidity, $\Delta\eta = |\eta_{jet1} - \eta_{jet2}|$ [12]. The DGLAP based generators are found

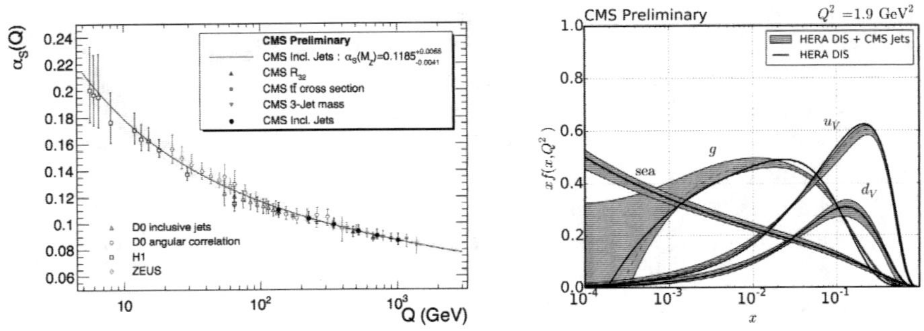

Figure 2. The running of the strong coupling $\alpha_s(Q)$ and its total uncertainty (left). Overview of the gluon, sea, u_V and d_V PDFs at starting scale $Q^2 = 1.9$ GeV^2 (right).

to be describing the complex observables well. But, no conclusion about the presence of BFKL like parton dynamics can be drawn from the present measurements.

Hadron production in pp collisions at $\sqrt{s} = 8$ TeV is measured by the CMS and TOTEM experiments [13]. An inclusive sample which events were triggered by forward activity in either side of the detector and a non-single-diffractive enhanced sample which events were triggered by both TOTEM T2 telescopes are used for the measurement. The production yields of leading charged-particle jets and charged particles [14] provide information on the underlying parton-parton cross section mechanism. A large sensitivity to the underlying physics is seen in the comparisons of data to various MC event generators. The measurement indicates the saturation of parton-parton cross sections can be seen already in a few GeV of p_T range at LHC energies.

3. Summary

An extensive list of QCD studies is presented from the CMS experiment. The large pp collision dataset has been collected during the Run I Phase and the CMS collaboration has explored measurements that shed light on a new era. Additionally the α_s has been extracted from jet measurements up to the TeV scale for the first time.

Acknowledgments

The author would like to acknowledge the Adiyaman University Scientific Activity Foundation for the funding.

References

[1] S. Chatrchyan et al. [CMS Collaboration], *JINST* **3**, S08004 (2008).
[2] S. Chatrchyan et al. [CMS Collaboration], *Phys. Rev.* D **87**, 112002 (2013).
[3] S. Chatrchyan et al. [CMS Collaboration], *Eur. Phys. J.* C **73** 2674 (2013).
[4] S. Chatrchyan et al. [CMS Collaboration], *JHEP* **06** 036 (2012).
[5] CMS Collaboration, *CMS PAS* FSQ-12-031 (2013).
[6] CMS Collaboration, *CMS PAS* SMP-12-012 (2013).
[7] CMS Collaboration, *CMS PAS* SMP-14-002 (2014).
[8] S. Chatrchyan et al. [CMS Collaboration], *Eur. Phys. J.* C **73** 2604 (2013).
[9] CMS Collaboration, *CMS PAS* SMP-12-027 (2013).
[10] CMS Collaboration, *CMS PAS* SMP-12-028 (2013).
[11] CMS Collaboration, *CMS PAS* FSQ-12-002 (2012).
[12] CMS Collaboration, *CMS PAS* FSQ-12-008 (2014).
[13] CMS Collaboration, *CMS PAS* FSQ-12-008 (2014).
[14] CMS Collaboration, *CMS PAS* FSQ-12-032 (2014).

Thermal behaviors of light unflavored tensor mesons in the framework of QCD sum rule

K. Azizi[†1], **A. Türkan**[*2], **H. Sundu**[*3], **E. Veli Veliev**[*4], **E. Yazıcı**[*5]

[†]Physics Division, Faculty of Arts and Sciences, Doğuş University
Acıbadem-Kadıköy, 34722 Istanbul, Turkey
[*]Department of Physics, Kocaeli University, 41380 Izmit, Turkey

E-mail: kazizi@dogus.edu.tr

Abstract. In this paper, we investigated the sensitivity of the masses and decay constants of $f_2(1270)$ and $a_2(1320)$ tensor mesons to the temperature using OCD sum rule approach. In our calculations, we take into account new additional operators appearing in operator product expansion (OPE). At the end of numerical analyses we show that at deconfinement temperature the decay constants and masses decreased by 6% and 96% of their vacuum values, respectively. Our results on the masses and decay constants at zero temperature of the tensor mesons are consistent with the experimental data as well as the vacuum sum rules predictions.

1. Introduction

Recent years, in order to understand properties of matter at high energies heavy-ion collision experiments are performed [1-6]. These experiments have a critical role on investigating hadron structures. Up to now, there were not enough experimental results on quark-gluon plasma (QGP), but with increasing energies on heavy-ion colliders, theoretical studies on the strong interaction at finite temperature have become more important. According to thermal QCD, around critical temperature, $T_c = 175MeV$, a transition occurs from hadronic matter to QGP phase which probably exists in early universe and neutron stars. Hence, theoretical efforts on thermal QCD calculations are important to understand the phase diagram and other properties of strong interactions [7-24].

In order to investigate Chiral symetry at high temperatures one needs to modify hadronic parameters for medium. There are many works on the medium corrections of hadron parameters by using different approaches such as chiral model [17,19], coupled channel [19] and sum rules [20-23] approaches. But still the thermal behaviors of tensor mesons are not known well. In this study, we use the thermal QCD sum rule method which is one of the most efficient tools [12,13] to calculate hadronic parameters of light unflavored tensor mesons. According to this method, hadron parameters are evaluated by using interpolating currents and operator product expansion approach (OPE). To extend the QCD sum rule method to finite temperatures, we assume that the Wilson expansion and the quark-hadron duality approximation are valid, but the quark and gluon condensates are changed with thermal expectation expressions [13]. In thermal QCD sum rule, the Lorentz invariance is broken and in the Wilson expansion appear extra operators which are expressed in terms of 4-vector velocity of the media and the energy-momentum tensor

[14-16]. Taking into account these new additional operators at finite temperature sum rule is obtained.

2. QCD sum rule for unflavored light mesons

In this section our aim is to obtain spectral density and non-perturbative part of correlation function to show how the physical quantities changes with the temperature. To calculate the spectral density we need to compute two point correlation function which is given as

$$\Pi_{\mu\nu,\alpha\beta}(q,T) = i\int d^4x e^{iq\cdot(x-y)}\langle\mathcal{T}[J_{\mu\nu}(x)\bar{J}_{\alpha\beta}(y)]\rangle\mid_{y=0}, \tag{1}$$

where $J_{\mu\nu}$ is the interpolating current of the tensor mesons as the following form:

$$\begin{aligned} J^{f_2(a_2)}_{\mu\nu}(x) &= \frac{i}{2\sqrt{2}}\Big[\bar{u}(x)\gamma_\mu D_\nu(x)u(x) + \bar{u}(x)\gamma_\nu D_\mu(x)u(x) \\ &\pm \bar{d}(x)\gamma_\mu D_\nu(x)d(x) \pm \bar{d}(x)\gamma_\nu D_\mu(x)d(x)\Big]. \end{aligned} \tag{2}$$

Here terms have positive sign for the f_2 meson and terms have negative sign for the a_2 meson in our calculations. Also $D_\mu(x)$ contain four-derivatives, acting on the left and right, simultaneously, with respect to the space-time. After applying derivatives we will put $y = 0$. Using the free quark propagator for light quarks we obtained the spectral density in the following form:

$$\rho_{f_2(a_2)}(s) = \frac{(m_u^2 + m_d^2)s}{32\pi^2} + \frac{3s^2}{160\pi^2}. \tag{3}$$

Using the non-perturbative parts of quark propagator as follows

$$\begin{aligned} S_q^{ij}(x-y) &= -\frac{\langle\bar{q}q\rangle}{12}\delta_{ij} - \frac{(x-y)^2}{192}m_0^2\langle\bar{q}q\rangle\Big[1 - i\frac{m_q}{6}(\not{x}-\not{y})\Big]\delta_{ij} \\ &+ \frac{i}{3}\Big[(\not{x}-\not{y})\Big(\frac{m_q}{16}\langle\bar{q}q\rangle - \frac{1}{12}\langle u\Theta^f u\rangle\Big) + \frac{1}{3}\Big(u\cdot(x-y)\not{u}\langle u\Theta^f u\rangle\Big)\Big]\delta_{ij}, \end{aligned} \tag{4}$$

we obtain the non-perturbative contributions which contain additional operators arising at finite temperature. $\Theta^f_{\mu\nu}$ is the fermionic part of the energy momentum tensor and u_μ is the four-velocity of the heat bath in Eq. (4). After matching the hadronic and OPE representations of correlator thermal sum rules for the decay constants obtained as:

$$\frac{f^2_{f_2(a_2)}(T)m^6_{f_2(a_2)}(T)}{m^2_{f_2(a_2)} - q^2} = \int_{(m_u+m_d)^2}^{s_0(T)} ds\frac{\rho_{f_2(a_2)}(s)}{s-q^2} + \Pi^{non-pert}_{f_2(a_2)}, \tag{5}$$

where non-perturbative contributions can be written as

$$\Pi^{non-pert}_{f_2(a_2)} = \frac{m_d m_0^2}{144q^2}\langle\bar{d}d\rangle + \frac{m_u m_0^2}{144q^2}\langle\bar{u}u\rangle - \frac{2\langle u\Theta^f u\rangle(q\cdot u)^2}{9q^2}. \tag{6}$$

3. Conclusions and Discussions

Last step is to determine the working region of two parameters: Borel mass parameter M^2 and the hadronic threshold at zero temperature $s_0(T=0)$. Determination of these parameters is important due to sum rules results should be stable under their small variations. In this work, we have chosen the intervals $s_0 = (2.2-2.5)GeV^2$ and $s_0 = (2.4-2.7)GeV^2$ for the continuum thresholds in f_2 and a_2 cases, respectively. Also we have chosen the working region of Borel mass as $1.4GeV^2 \le M^2 \le 3GeV^2$.

Taking into account the temperature dependencies of hadronic threshold, energy density and quark condensates, we obtained that masses and decay constants are well described by the following fit functions:

$$m_{f_2} = -1.0549 \times 10^{-5} e^{72.6744T} + 1.26519, \tag{7}$$

$$f_{f_2} = -4.28022 \times 10^{-6} e^{42.8082T} + 0.04258, \tag{8}$$

and

$$m_{a_2} = -1.25459 \times 10^{-5} e^{71.8390T} + 1.3221, \tag{9}$$

$$f_{a_2} = -4.06115 \times 10^{-6} e^{42.6621T} + 0.04232. \tag{10}$$

Here decay constants, masses and temperature are expressed in units of GeV. These parameterizations are valid only in the interval $0 \le T \le 0.16GeV$.

Our investigations show that the values of masses and decay constants are stable until temperature 0.1 GeV but after this point they start to decrease with altering the temperature. Also at deconfinement temperature, the decay constants and masses decreased by 6% and 96% of their vacuum values, respectively

Our results show that at zero temperature the masses and decay constants are $m_{f_2} = 1.28 \pm 0.08GeV$, $f_{f_2} = 0.041 \pm 0.002$, $m_{a_2} = 1.33 \pm 0.10GeV$ and $f_{a_2} = 0.042 \pm 0.002$. Our results for masses are compatible with the vacuum sum rules predictions. Our predictions on the temperature behaviors of decay constants and masses can be verified in the future experiments.

This work has been supported in part by the Scientific and Technological Research Council of Turkey (TUBITAK) under the research projects 110T284 and 114F018.

References

[1] Apel W D et al. (Serpukhov-CERN Collaboration) 1975 *Phys. Lett.* B **57** 398
[2] Longacre R S et al. 1986 *Phys. Lett.* B **177** 223
[3] Doser M et al. (ASTERIX Collaboration) 1988 *Phys. Lett.* B **215** 792
[4] Kubota Y et al. (CLEO Collaboration) 1994 *Phys. Rev. Lett.* **72** 1972
[5] Ablikim M et al. (BES Collaboration) 2005 *Phys. Lett.* B **607** 243
[6] Aaij R et al. (LHCb Collaboration) 2011 *Phys. Lett.* B **698** 14
[7] Aliev T M and Shifman M A 1982 *Phys. Lett.* B **112** 401
[8] Ebert D, Faustov R N and Galkin V O 2009 *Phys. Rev.* D **79** 114029
[9] Aliev T M, Azizi K and Bashiry V 2010 *J. Phys.* G **37** 025001
[10] Aliev T M, Azizi K, Savcı M *Phys.Lett.* B **690** 164
[11] Sundu H and Azizi K 2012 *Eur. Phys. J.* A **48** 81
[12] Shifman M A, Vainshtein A I and Zakharov V I 1979 *Nucl. Phys.* B **147** 385; 1979 *Nucl. Phys.* B **147** 448
[13] Bochkarev A I and Shaposhnikov M E 1986 *Nucl. Phys.* B **268** 220
[14] Hatsuda T, Koike Y and Lee S H 1993 *Nucl. Phys.* B **394** 221
[15] Mallik S 1998 *Phys. Lett.* B **416** 373
[16] Shuryak E V 1993 *Rev. Mod. Phys.* **65** 1
[17] Mallik S and Sarkar S 2002 *Eur. Phys. J.* C **25** 445
[18] Mishra A, Bratkovskaya E L, Schaffner-Bielich J, Schramm S and Stocker H 2004 *Phys. Rev.* C **69** 015202; 2004 *Phys. Rev.* C **70** 044904; Kumar A and Mishra A 2010 *Phys, Rev.* C **81** 065204
[19] Waas T, Kaiser N, Weise W 1996 *Phys. Lett.* B **365** 12
[20] Veliev E V 2008 *J. Phys.* G **35** 035004; Veliev E V and Aliev T M 2008 *J. Phys.* G **35** 125002
[21] Veliev E V, Azizi K, Sundu H and Aksit N 2012 *J. Phys.* G **39** 015002; Veliev E V and Kaya G 2009 *Eur. Phys. J.* C **63** 87
[22] Dominguez C A, Loewe M, Rojas J C and Zhang Y 2010 *Phys. Rev.* D **81** 014007; Dominguez C A, Loewe M and Rojas J C 2007 *JHEP* **08** 040
[23] Morita K and Lee S H 2008 *Phys. Rev.* C **77** 064904; 2010 *Phys. Rev.* D **82** 054008
[24] Veliev E V, Azizi K, Sundu H, Kaya G and Türkan A 2011 *Eur. Phys. J.* A **47** 110

AUTHOR INDEX

Institute of Physics
Dirac House, Temple Back
Bristol BS1 6BE UK

CURRAN ASSOCIATES INC
proceedings.com